Python for Mathematics

Python for Mathematics introduces readers to effective methods for doing mathematics using the Python programming language. Most programming texts introduce readers to the building blocks of programming and build up to using more sophisticated tools for a specific purpose, like doing particular mathematical tasks. This is akin to teaching someone how to forge metal so as to make a nail, and then slowly working up to using sophisticated power tools so as actually to build something. This book does things in a different way, by first getting readers to begin using and understanding the tools that are going to be helpful to them as mathematicians, and only then moving on to the granular details. In this way, the practical application of the tools can aid in the understanding of the theory.

Features
- Complete with engaging, practical exercises
- Many useful and detailed coding examples
- Suitable for undergraduates in mathematics, and other quantitative sciences
- Empowers readers to design and create their own Python tools.

Vincent Knight is Professor of Mathematics at Cardiff University in the School of Mathematics. His research interests are in emergent behaviour, probabilistic modelling, applications in healthcare, and pedagogy. He maintains a number of open-source research software projects, has been a trustee of the UK Python association, is an editor for the Journal of Open-Source Software, was awarded the 2017 John Pinner award for contribution to the Python community, and is a fellow of the Sustainable Software Institute. He regularly wins awards for his teaching at the School of Mathematics. He does not only speak at conferences around the world but continues to organise conferences to bring the power of open-source software to as many people as possible.

Chapman & Hall/CRC The Python Series

Python has been ranked as the most popular programming language, and it is widely used in education and industry. This book series will offer a wide range of books on Python for students and professionals. Titles in the series will help users learn the language at an introductory and advanced level, and explore its many applications in data science, AI, and machine learning. Series titles can also be supplemented with Jupyter notebooks.

Python Packages
Tomas Beuzen and Tiffany-Anne Timbers

Statistics and Data Visualisation with Python
Jesús Rogel-Salazar

Introduction to Python for Humanists
William J.B. Mattingly

Python for Scientific Computation and Artificial Intelligence
Stephen Lynch

Learning Professional Python Volume 1: The Basics
Usharani Bhimavarapu and Jude D. Hemanth

Learning Professional Python Volume 2: Advanced
Usharani Bhimavarapu and Jude D. Hemanth

Learning Advanced Python from Open Source Projects
Rongpeng Li

Foundations of Data Science with Python
John Mark Shea

Data Mining with Python: Theory, Applications, and Case Studies
Di Wu

A Simple Introduction to Python
Stephen Lynch

Introduction to Python: with Applications in Optimization, Image and Video Processing, and Machine Learning
David Baez-Lopez and David Alfredo Báez Villegas

Tidy Finance with Python
Christoph Frey, Christoph Scheuch, Stefan Voigt, and Patrick Weiss

Introduction to Quantitative Social Science with Python
Weiqi Zhang and Dmitry Zinoviev

Python Programming for Mathematics
Julien Guillod

Geocomputation with Python
Michael Dorman, Anita Graser, Jakub Nowosad, and Robin Lovelace

Python for Mathematics
Vincent Knight

BiteSize Python for Absolute Beginners: With Practice Labs, Real-World Examples, and Generative AI Assistance
Di Wu

For more information about this series please visit: https://www.routledge.com/Chapman--HallCRC-The-Python-Series/book-series/PYTH

Python for Mathematics

Vincent Knight

CRC Press

Taylor & Francis Group

Boca Raton London New York

CRC Press is an imprint of the
Taylor & Francis Group, an **informa** business

A CHAPMAN & HALL BOOK

Front cover image: Mendin\Shutterstock

First edition published 2025
by CRC Press
2385 NW Executive Center Drive, Suite 320, Boca Raton FL 33431

and by CRC Press
4 Park Square, Milton Park, Abingdon, Oxon, OX14 4RN

CRC Press is an imprint of Taylor & Francis Group, LLC

© 2025 Vincent Knight

ISBN: 978-1-032-58817-9 (hbk)
ISBN: 978-1-032-58218-4 (pbk)
ISBN: 978-1-003-45186-0 (ebk)

DOI: 10.1201/9781003451860

Typeset in Latin Modern font
by KnowledgeWorks Global Ltd.

Publisher's note: This book has been prepared from camera-ready copy provided by the authors.

To Daniele, may someone who reads this book learn a fraction of what I have learnt from you. I am grateful for our friendship.

Contents

CHAPTER   17 ▪ Testing   227

SECTION IV About This Book

CHAPTER   18 ▪ How This Book Is Written   247

Preface

Welcome to this book.

This is not a book for learning to program to do mathematics. There are many excellent books that do this [1, 6, 10]. This is a book for people who would like to learn how to use programming tools to assist with when doing Mathematics.

Mathematics is often thought of as solving problems. In secondary school this can be sets of quadratic equations that need to be solved or probabilities of specific hands of cards that need to be calculated.

As one progresses further into mathematics, the subject becomes less about solving problems through mechanical calculation and more about using our mathematical knowledge and insight to choose **which problems to solve**.

This is what this book attempts to address. It aims to be a user guide for how the Python programming language can be used to reduce mechanical calculation which leaves more space to do real mathematics.

Whilst no book should ever try to stop a mathematician from picking up a pen and pencil and thinking about a problem, this one does aim to show how modern mathematicians can replace, some of, the use of their pen with openly available Python tools. For example, in Chapter 3, how to solve an equation by essentially just writing it down is covered. In Chapter 7 probabilities of specific events are simulated.

In the second part of this book, a more traditional approach of programming with Python is used to show how to build tools. Not only does this cover commonly taught programming techniques but also goes into principles of software development used in industry. For example, Chapter 16 covers a modern way of writing documentation for software and Chapter 17 covers how to write code that tests software.

This book is for you, whether you are a seasoned professional mathematician who would like to know some of the best practices for using Python or perhaps more typically, if you are a first year university student with an understanding of the mathematical topics covered. I hope you enjoy it.

I

Overview

CHAPTER 1

Introduction

1.1   WHO IS THIS BOOK FOR?

This book aims to introduce readers to programming **for** mathematics.

It is assumed that readers are used to solving secondary school mathematics problems of the form:

Given the function $f : \mathbb{R} \to \mathbb{R}$ defined by $f(x) = x^2 - 3x + 1$, obtain the global minima of the function.

To solve this, you need to apply **mathematical knowledge**:

1. Differentiate $f(x)$ to get $\frac{df}{dx}$;

2. Equate $\frac{df}{dx} = 0$;

3. Use the second derivative test on the solution to the previous equation.

For each of those three steps you will usually make use of our **mathematical techniques**:

1. Differentiate $f(x)$:

$$\frac{df}{dx} = 2x - 3$$

2. Equate $\frac{df}{dx} = 0$:

$$2x - 3 = 0 \Rightarrow x = 3/2$$

3. Use the second derivative test on the solution:

$$\frac{d^2 f}{dx^2} = 2 > 0 \text{ for all values of } x$$

Thus $x = 3/2$ is the global minima of the function.

As you progress as a mathematician, **mathematical knowledge** is more prominent than **mathematical technique**: often knowing what to do is the real problem as opposed to having the technical ability to do it.

This is what this book will cover: **programming** allows you to instruct a computer to carry out mathematical techniques.

DOI: 10.1201/9781003451860-1

For example, you will learn how to solve the above problem by instructing a computer which **mathematical technique** to carry out.

This book covers how to give the correct instructions to a computer.

The following is an example; do not worry too much about the specific code used for now:

Differentiate $f(x)$ to get $\frac{df}{dx}$

```
1  import sympy as sym
2
3  x = sym.Symbol("x")
4  sym.diff(x ** 2 - 3 * x + 1, x)
```

$$2x - 3$$

Equate $\frac{df}{dx} = 0$

```
1  sym.solveset(2 * x - 3, x)
```

$$\left\{ \frac{3}{2} \right\}$$

Use the second derivative test on the solution

```
1  sym.diff(x ** 2 - 3 * x + 1, x, 2)
```

$$2$$

Figure 1.1 shows a summary.

1.1.1 How is this book different from similar books?

A traditional structure of this book would probably be to re-order the chapters as follows:

1. Chapter 2

2. Chapter on variables (seen in Chapter 11)

3. Chapter on conditionals (seen in Chapter 11)

4. Chapter on loops (seen in Chapter 11)

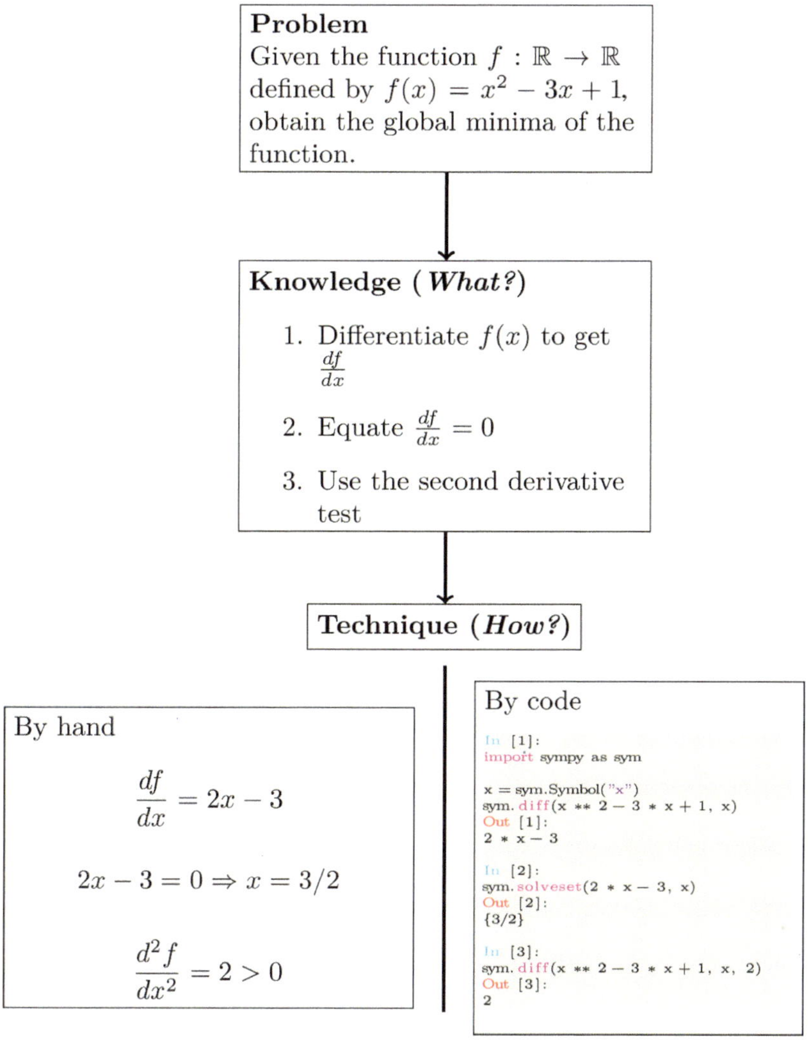

Problem
Given the function $f : \mathbb{R} \to \mathbb{R}$ defined by $f(x) = x^2 - 3x + 1$, obtain the global minima of the function.

Knowledge (*What?*)

1. Differentiate $f(x)$ to get $\frac{df}{dx}$

2. Equate $\frac{df}{dx} = 0$

3. Use the second derivative test

Technique (*How?*)

By hand

$$\frac{df}{dx} = 2x - 3$$

$$2x - 3 = 0 \Rightarrow x = 3/2$$

$$\frac{d^2 f}{dx^2} = 2 > 0$$

By code

```
In [1]:
import sympy as sym

x = sym.Symbol("x")
sym.diff(x ** 2 - 3 * x + 1, x)
Out [1]:
2 * x - 3

In [2]:
sym.solveset(2 * x - 3, x)
Out [2]:
{3/2}

In [3]:
sym.diff(x ** 2 - 3 * x + 1, x, 2)
Out [3]:
2
```

Figure 1.1   Knowledge versus technique in this book.

5. Chapter on functions (seen in Chapter 12)

6. Chapters on data structures (seen in Chapter 12)

7. Chapter 13

8. Chapter on Sympy (with an overview of the topics in Chapters 3–5 and 10)

9. Chapter 6

10. Chapter 7

11. Chapter 8

12. Chapter 9

13. Chapter 14

14. Chapter 15

15. Chapter 16

16. Chapter 17

The choice to *flip* this structure and start with real use cases (and not code recipes) is deliberate. The tools covered in Chapters 3–10 can be used with little to no programming knowledge and need only an understanding of the mathematics. Following this, the topics covered in Chapters 11–13 let the reader expand on the knowledge and learn basics of programming. The topics and techniques covered in Chapters 14–17 show how modern research software is designed.

1.2   WHAT IS IN THIS BOOK?

Most programming texts introduce readers to the building blocks of programming and build up to using more sophisticated tools for a specific purpose.

This is akin to teaching someone how to forge metal so as to make a nail and then slowly work up to using more sophisticated tools such as power tools to build a house.

This book will do things in a different way: you will start with using and understanding tools that are helpful to mathematicians. In the later part of the book, you will cover the building blocks and you will be able to build your own sophisticated tools.

1.2.1   How is this book organised?

The book is in two parts:

1. Tools for mathematics and

2. Building tools.

The first part of the book will not make use of any novel mathematics. Instead you will consider a number of mathematics topics that are often covered in secondary school.

- Algebraic manipulation

- Calculus (differentiation and integration)

- Combinatorics (permutations and combinations)

- Probability

- Linear algebra

- Sequences

- Statistics

- Differential equations

The questions you will tackle aim to be familiar in their presentation and description. **What will be different** is that no **by hand** calculations will be done. You will instead carry them all out using a programming language.

In the second part of the book, you will be encouraged to build your own tools for tackling problems of your choice.

Each chapter will have four parts:

- A tutorial: you will be walked through solving a problem. You will be specifically told what to do and what to expect.

- A how to section: this will be a shorter more succinct section that will detail how to carry out specific things.

- A further information section: this will be a section with references to further resources as well as background information about specific things in the chapter and answers to common questions.

- An exercise section: this will be a number of exercises that you can work on.

There are a number of different Python libraries, programming techniques and frameworks covered in this book:

- Diátaxis (Chapter 16)

- Recursion (Chapter 8);

- `itertools` (Chapter 6)

- `random` (Chapters 7)

- `statistics` (Chapters 9)

- `sympy` (Chapters 3–5, 10);

1.2.2   How to use this book?

Readers are welcome to use this book in any way they find useful; however, it is designed with the following suggestions in mind:

- Start by following along with the tutorial. Carrying out the steps and observing the outcomes. It is not expected that a reader gains a deep understanding of a given topic when working through the tutorial. The goal here is to achieve some level of familiarity.

- After the tutorial, work through the how to section. It is through this section that a deeper understanding is to be gained by making connections to steps taken in the tutorial. **After working through the how to section it is hoped that the reader would understand all steps taken in the tutorial**.

- The exercise section is an opportunity for the reader to practice the topics in the how to section.

- After working through those three sections, it is possible that some readers have further questions or would like to find more information about a given topic. This is covered in the further information section.

1.2.3   How is code displayed in this book?

In this book, you will see code displayed in a number of different formats. The most common is the following:

Jupyter input

```
1    2 + 2
```

4

This is shown as input to the programming tool "Jupyter" which is described at length in Chapter 2. As well as the input, it will also display the output (as above).

You will see typical usage instructions for particular code commands:

Usage

```
1    sym.solveset(<equation>)
```

You will see how to write a particular language called "markdown" (covered in Chapters 2 and 16):

Markdown input

```
1    # Algebra with Python
```

In Chapters 14–17 you will also see Python code saved to Python files:

Python file

```
1    print(2 + 2)
```

You will also see commands written for a command line tool. This is how you will start "Jupyter" in Chapter 2 but will be introduced more formally in Chapter 14.

Command line input

```
1    $ ls
```

Note that when some lines of code are long, they might include a carriage return symbol ($\hookrightarrow$):

Jupyter input

```
1      123456789 + 123456789 + 123456789 + 123456789 + 123456789 +
     ↪   123456789 + 123456789
```

The carriage return symbol should be ignored as it is only present in the book due to the physical constraint of the page width. The code should still be written as a single line.

1.3   WHAT IS **NOT** THIS BOOK?

With thanks to the progressive understanding of the publisher, there is an online version of this book. As such, there are two specific things that are not in this book but are available in the online version:

1. Solutions to the exercises;

2. A collection of further information chapters; this covers specific tools like numpy for numerical mathematics as well as a more detailed description of working with Jupyter kernels.

As well as those two things, writing fixes, more exercises and new further information chapters will continue to be added to the online version.

II

Tools for Mathematics

CHAPTER 2

Using Notebooks

2.1 INTRODUCTION

At the advent of Calculus two mathematicians are credited with its formalisation/invention:

- Isaac Newton

- Gottfried Leibniz

One of the differences between the approaches taken by Newton and Leibniz is their notation. Newton denoted the derivative of a function f as:

$$Df$$

Leibniz denoted the derivative with the now more commonly used notation:

$$\frac{df}{dx}$$

The mathematics itself is unchanged: what changes is the language/notation used to communicate it. Similarly when giving instructions through code to a computer there are a number of notations, more commonly called languages available. This book will be using a language called **Python**. Python was originally designed as a teaching language but it is now popular both in academia and in industry.

In this chapter you will cover:

- Installing the specific distribution of Python on your computer.

- Using something called a Jupyter notebook to write and run Python code.

- Writing descriptive notes using markdown and LaTeX (pronounced Lay-tech).

2.2 TUTORIAL

This tutorial will take the reader through an example of using Jupyter notebooks. Jupyter is the interface to the Python programming language used in the first part of this book.

2.2.1 Installation

1. Navigate to https://www.anaconda.com/products/individual.

DOI: 10.1201/9781003451860-2

2. Identify and download the version of Python 3 for your operating system (Windows, MacOS, Linux).

3. Run the installer. I recommend using the default choices during the installation process.

If you have already used Python, it is still recommended that you use the Anaconda distribution. An explanation for this is available in Section 2.5.1. If you are using a Chromebook or a Tablet, there are websites and applications that you can use instead of Anaconda to follow along in this book. See the online version of this book for up-to-date information on this.

2.2.2 Starting a Jupyter notebook server

Open a command line tool:

1. On **MacOS** search for `terminal`. See Figure 2.1.

2. On **Windows** search for `Anaconda Prompt` (it should be available to you after installing Anaconda). See Figure 2.2.

In the command line tool type (without the $):

Command line input

```
1   $ jupyter notebook
```

Press `Enter` on your keyboard.

Throughout this book, when there are commands to be typed in a command line they will be prefixed them with a $. Do not type the $.

This will open a new page in your browser. The url bar at the top should have something that looks like: `http://localhost:8888/tree`. This is the general interface to the Jupyter server. It shows the general file structure on your computer as shown in Figure 2.3.

2.2.3 Creating a new notebook

In the top right, click on the `New` button (Figure 2.4) and click on `Notebook`. This will be followed by a prompt to choose the programming language to use, this is referred to as the kernel: select Python 3. Change the name of the notebook by clicking on "Untitled" and changing the name. You will call it "introduction" as shown in Figure 2.5.

2.2.4 Organising your files

Open your file browser:

1. File Explorer on **Windows** (see Figure 2.7)

2. Finder on **MacOS** (see Figure 2.6)

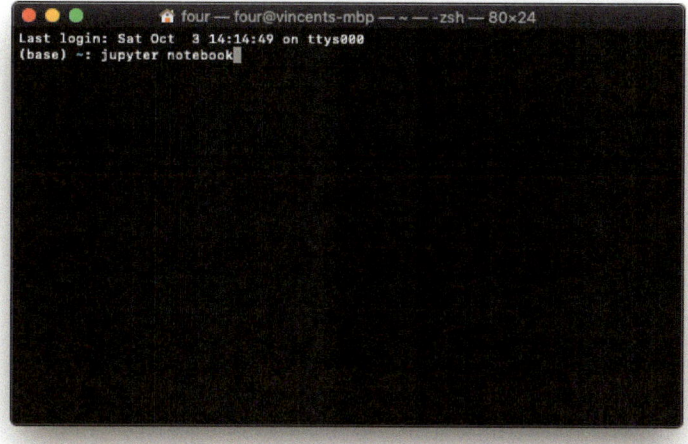

Figure 2.1   Starting the notebook server on MacOS.

Figure 2.2   Starting the notebook server on Windows.

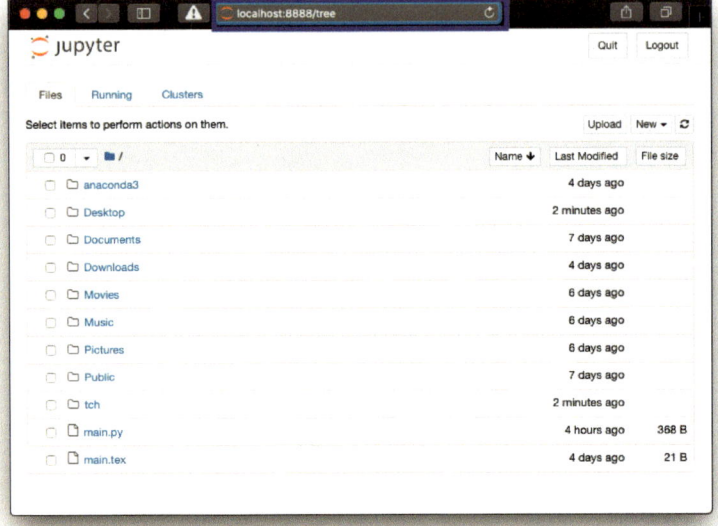

Figure 2.3 The Jupyter notebook interface.

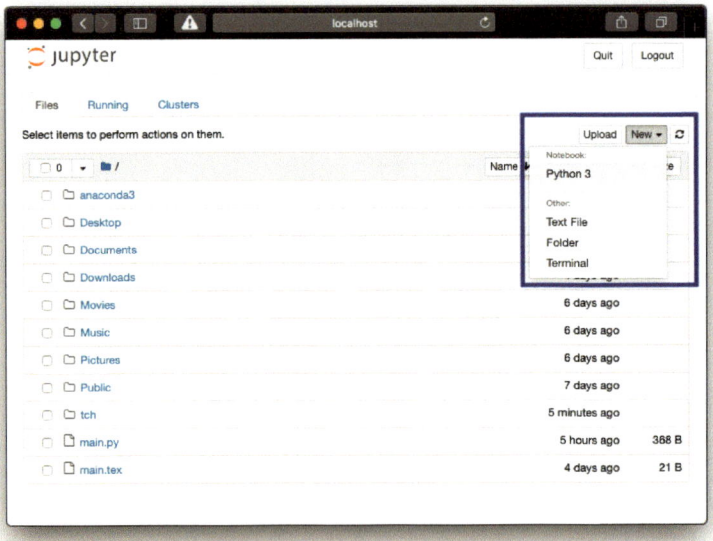

Figure 2.4 Creating a new notebook.

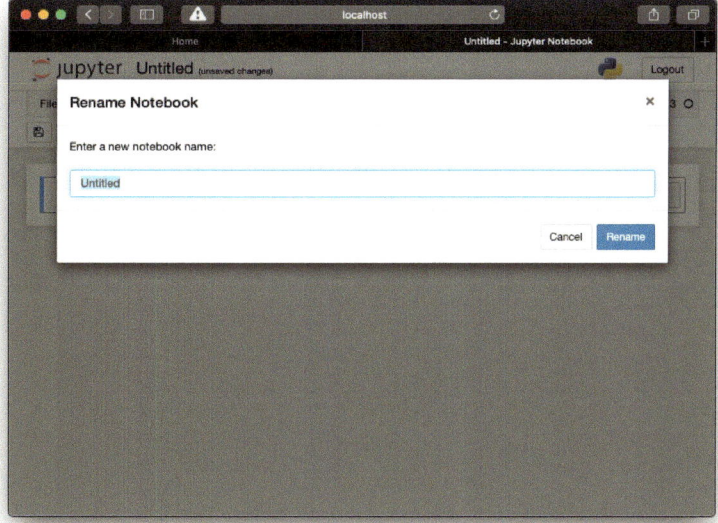

Figure 2.5   Changing the notebook name.

Navigate to where your notebook is (this might not be immediately evident): you should see a `introduction.ipynb` file. Find a location on your computer where you want to keep the files for this book, using your file browser:

1. Create a new directory called `pfm` (short for "Python for Mathematics");

2. Inside that directory create a new directory called `nbs` (short for "Notebooks");

3. Move the `introduction.ipynb` file to this `nbs` directory.

Figure 2.6   Creating a new directory on MacOS.

2.2.5   Writing some basic Python code

Go back to the Jupyter notebook server (in your browser). Use the interface to navigate to the `pfm` directory and inside that the `nbs` directory and open the `introduction.ipynb` notebook.

Figure 2.7   Creating a new directory on Windows.

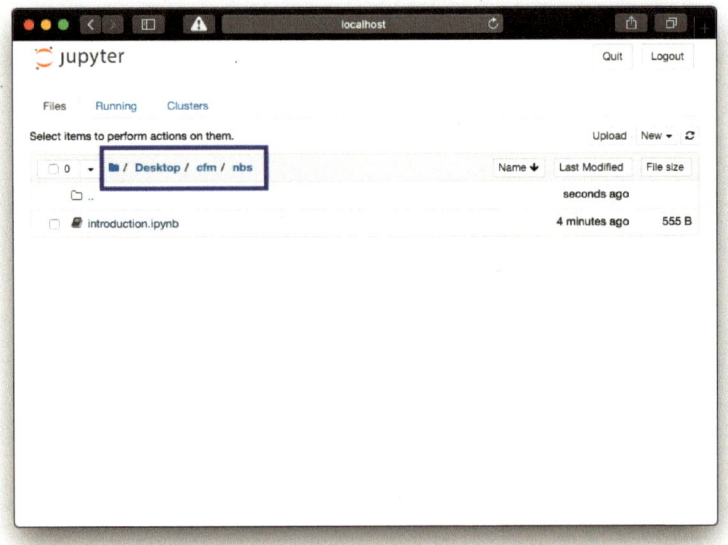

Figure 2.8   Opening a notebook.

In the first available "cell" write the following calculation:

Usage

```
1       2 + 2
```

When you have done that click on the `Run` button shown in Figure 2.9. You can also use `Shift + Enter` as a keyboard shortcut.

Jupyter input

```
1       2 + 2
```

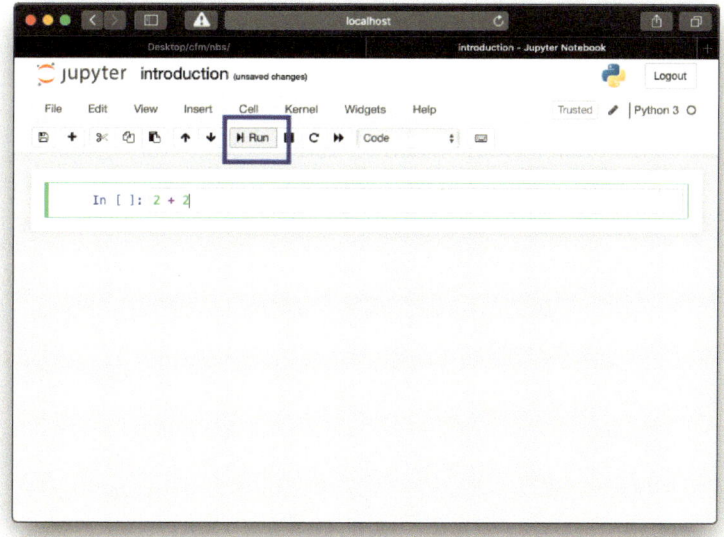

Figure 2.9   Running code.

2

Figure 2.9 shows two different things:

1. The input: which is the instruction to Python to use the mathematical technique of addition to compute $2 + 2$.

2. The output: showing the output that Python has returned as a result of the instruction.

2.2.6 Writing markdown

One of the reasons for using Jupyter notebooks is that it allows a user to include both code and communication using something called `markdown`. Create a new cell and change the cell type to `Markdown`. Now write the following in there:

Markdown input

```
1  As well as using Python in Jupyter notebooks you can also write using
2  Markdown. This allows us to use basic \LaTeX\; as a way to display
3  mathematics. For example:
4
5  1. $\frac{2}{3}$
6  2. $\sum_{i=0}^n i$
```

When you run that it should look like Figure 2.10.

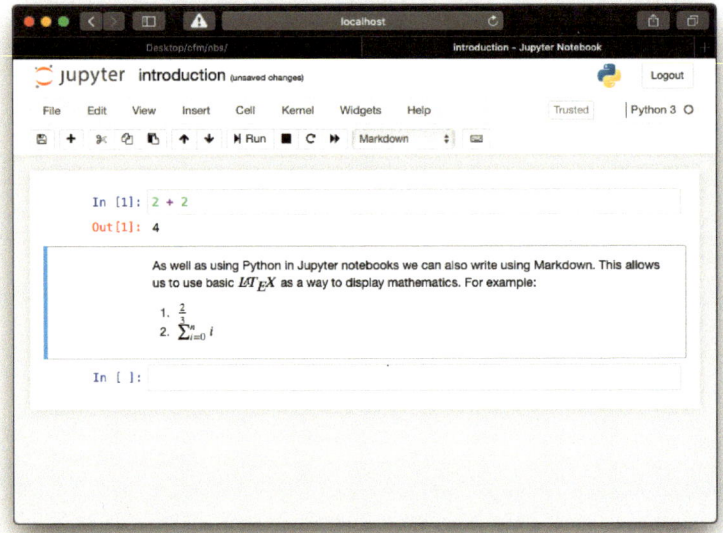

Figure 2.10   Rendering markdown.

2.2.7 Saving your notebook to a different format

Click on `File` and `Download As` which brings up a number of formats that Jupyter notebooks can be exported to. Some of these might need other tools installed on your computer but a portable option is HTML. Click on `HTML (.html)`. Now use your file browser and open the downloaded file. This will open in your browser a static version of the file you have been working on. This is a helpful way to share your work with someone who might not have Jupyter (or even Python).

This tutorial has:

- Installed the Anaconda distribution of Python.

- Started a notebook server.

- Created a new notebook.

- Run some Python code.

- Written some markdown.

- Saved the notebook to a different format.

2.3 HOW TO

2.3.1 Install Anaconda

1. Navigate to `https://www.anaconda.com/download/success`

2. Download the distribution of anaconda for your Operating System

3. Run the installer

2.3.2 Start a Jupyter notebook server

1. Open a command line tool (`Anaconda prompt`) on Windows, `terminal` on OS X);

2. Type `jupyter notebook` and press `Enter`

2.3.3 Create a new notebook

1. Navigate to the location you want using the Jupyter interface;

2. Click on the `new` button in the top right;

3. Rename the notebook (change `Untitled` in the top left to a name of your choice).

2.3.4 Find/open a notebook

Using a file browser you can navigate the directories and files on your computer. Jupyter notebooks appear as generic files with the `.ipynb` extension.

You cannot double click on these to open them, you need to navigate to them through the Jupyter interface.

2.3.5 Run Python code

In a Jupyter notebook cell write an instruction, for example:

```
Usage
1   3 / 5
```

and click on the `Run` button or use `Shift + Enter` as a keyboard shortcut.

2.3.6   Carry out basic arithmetic operations

The Python code for the following arithmetic operations are:

1. Addition, $2 + 2$: `2 + 2`;

2. Subtraction, $3 - 1$: `3 - 1`;

3. Multiplication, 3×5: `3 * 5`;

4. Division, $20/5$: `20 / 5`;

5. Exponentiation, 2^4: `2 ** 4`;

6. Integer remainder, $5 \mod 2$: `5 % 2`;

7. Combining operations, $\frac{2^3+1}{4}$: `(2 ** 3 + 1) / 4`;

   Note that instructions to a computer (through the code you write) need to be specific. For example, the ^ symbol in Python does not mean exponentiation. If you were to type 2 ^ 4, you would get an error. In later chapters you will see what the specific instructions are to carry out more complex operations.

2.3.7   Write markdown

To write markdown click on a cell and change the type to `Markdown`, you can do this by clicking on `Cell`, `Cell Type` or by using the scroll wheel in the menu bar. Markdown is a lightweight mark up language that allows you to write and include various types of formatting which include:

1. Headings;

2. Bold and italics;

3. Ordered and unordered lists;

4. Code (which will only be displayed but not run);

5. Hyperlinks

   A more detailed guide for writing markdown is given in Section 16.2.2.

2.3.8   Write basic LaTeX

Jupyter notebooks allow for markdown cells to not only include markdown but also include mathematics using another mark up language called LATEX. Here is a brief overview of the syntax for arithmetic operations:

- `$a+b$` gives: $a + b$:

- `$a-b$` gives: $a - b$

- `$-a$` gives: $-a$

- `$ab$` gives ab

- `$a\cdot b$` gives $a \cdot b$

- $a\times b$ gives $a \times b$

- a/b gives a/b

- $\frac{a}{b}$ gives $\frac{a}{b}$

- $a \char94 b$ gives a^b

The $<expression>$ delimiters create what is called an "inline" mathematics environment. You can change these to $$<expression>$$ to give "displayed mathematics".

You can write a matrix:

```
\begin{pmatrix}
    a&b\\
    c&d\\
    e&f\\
\end{pmatrix}
```

gives:

$$
\begin{pmatrix}
a & b \\
c & d \\
e & f
\end{pmatrix}
$$

You can write integrals:

```
$$
    \int_{0}^{\infty}x dx
$$
```

gives:

$$
\int_{0}^{\infty} x\,dx
$$

You can write summations:

```
$$
    \sum_{0}^{n}i
$$
```

gives:

$$
\sum_{0}^{n} i
$$

2.3.9 Save the output in a different format

Click on File, then Download as and choose the format you want to use. HTML is a portable option that can be viewed on most devices, note however that you cannot run the cells: what you are downloading is a static version of your notebook.

2.4 EXERCISES

1. Create a new notebook rename it "exercises". Navigate to it using your file browser to make sure you can find it.

2. Write and run some Python code to carry out the following calculations.

    (a) $3 + 8$

    (b) $3/7$

    (c) $456/21$

    (d) $\frac{4^3+2}{2\times 5} - 5^{\frac{1}{2}}$

3. Write a markdown cell with the following and view the rendered version:

    ```
    # Euler's equation

    $$
    e ^ {i\pi} = -1
    $$
    ```

4. Render the following expressions by writing markdown.

    (a) $\frac{4^3+2}{2\times 5}$

    (b) $-5^{\frac{1}{2}}$

    (c) $\frac{df}{dx}$

    (d) $\int_5^{12} x^2 dx$

    (e) $\begin{pmatrix} 4 & 12 & 3 \\ 2 & x & i \end{pmatrix}$

5. Save your notebook to HTML and open and view it.

6. Download the notebook available at 10.5281/zenodo.7118738 and check that you are able to open it.

2.5 FURTHER INFORMATION

2.5.1 Why use anaconda?

Python is a free and open source piece of software. One of the main reasons for its popularity is that there are a number of separate tools that work well with it, these are called libraries. Sometimes installing these libraries can require an understanding of some potential pitfalls. In scientific circles the Anaconda distribution was developed to give a single download of not only Python but a lot of commonly used libraries.

2.5.2 Why use Jupyter?

There are variety of ways to write and run Python:

1. Using an interactive notebook environment like Jupyter;

2. Using an integrated development environment and/or editor.

The second part of this book will use an editor. One strength of Jupyter is that it allows you to include communication (writing through markdown) with your code. This allows you to use code and describe what you are using it for. Another advantage is that it allows you to immediately have your output next to your input. There are some limitations to Jupyter as an editor which is why you will explore using a powerful editor in the second part of the course. In general:

1. Jupyter is a fantastic way to interactively use and communicate code;

2. Integrated development environments and/or editors are the correct tool to write software.

In this book you will learn to use either approach in the appropriate manner for the right task. For the first part, code will be used interactively and so you will use Jupyter notebooks.

2.5.3   Why can I not double click on a Jupyter notebook file?

When you double click on a file and your computer opens it in an application because a default is set for the particular file extension. For example, double clicking on `main.docx` will automatically open up the document using a word processor (like Microsoft word). This is because the file has the extension `.docx` and your operating system has set that anything with that extension will be opened in that particular application. You could also open the application and navigate to the file and open it directly.

With Jupyter notebooks no default is set by the operating system as the application that opens it is in fact a local web server in your browser. As such you do not have a choice and need to open it in the Jupyter interface.

2.5.4   Where can I find keyboard shortcuts for using Jupyter

In a notebook if you go to the menu bar and click on `Help` followed by `Keyboard Shortcuts` you will find a number of helpful keyboard shortcuts.

For example, when on a cell pressing `Esc` followed by `m` will turn the cell into a markdown cell.

2.5.5   What is markdown?

As described at `https://www.markdownguide.org/getting-started/`:

> "Markdown is a lightweight markup language that you can use to add formatting elements to plain text documents. Created by John Gruber in 2004, Markdown is now one of the world's most popular markup languages."

2.5.6   What is LaTeX?

As described at `https://www.latex-project.org/about/`:

> "LaTeX;, which is pronounced 'Lah-tech' or 'Lay-tech' (to rhyme with 'blech' or 'Bertolt Brecht'), is a document preparation system for high-quality typesetting. It is most often used for medium-to-large technical or scientific documents but it can be used for almost any form of publishing."

"LaTeX is not a word processor! Instead, LaTeX encourages authors not to worry too much about the appearance of their documents but to concentrate on getting the right content."

2.5.7 Can I use \(and \) instead of $ for LATEX?

You will see in some places that \(, \) or \[, \] can be used as delimiters for LATEX when used outside of Jupyter notebooks. This is in fact recommended for a number of reasons, one of which is given at `https://vknight.org/tex/`:

"Note that using \(and \) is preferred over $. One of the reasons is that it is easier for humans (and machines) to find the start and end of some mathematics."

> If you want to use \(, \) or \[, \] as mathematics delimiters within Jupyter notebooks, you need to escape the \ and use: \\(, \\) or \\[, \\] instead.

2.5.8 What is a markup language?

LATEX and markdown are both examples of what is called a **markup language**. Another common example of a markup language is html (the way web pages are written). A markup language is a system that allows us to write content alongside annotations to specify how the content is to appear. This description of markdown from `https://www.markdownguide.org/getting-started/` applies to any markup language:

"Using Markdown is different than using a WYSIWYG editor. In an application like Microsoft Word, you click buttons to format words and phrases, and the changes are visible immediately. Markdown isn't like that. When you create a Markdown-formatted file, you add Markdown syntax to the text to indicate which words and phrases should look different."

In general whilst it might take a little while to learn all the intricacies of a markup language it allows for more portability and precision. Markup languages differ in complexity:

- LATEX is incredibly sophisticated and has a huge range of capabilities.

- Markdown is designed to be basic with a few specific annotations to remember.

CHAPTER 3

Algebra

3.1 INTRODUCTION

A typical secondary school curriculum includes Algebra which is described in the A-level syllabus as:

"Algebra: this is an essential tool which supports and expresses mathematical reasoning and provides a means to generalise across a number of contexts."

In practice this often means:

- Manipulating numeric expressions;

- Manipulating symbolic expressions;

- Solving equations.

You can use a computer to carry out some of these techniques.

This chapter covers:

- Manipulating numeric and symbolic expressions.

- Solving equations.

3.2 TUTORIAL

To demonstrate the ways in which a computer can assist with Algebra, in this tutorial you will solve the following two problems:

1. Rationalise the denominator of $\frac{1}{\sqrt{2}+1}$

2. Consider the : $f(x) = 2x^2 + x + 1$:

    (a) Calculate the discriminant of the equation $2x^2 + x + 1 = 0$. What does this tell you about the solutions to the equation? What does this tell you about the graph of $f(x)$?

    (b) By completing the square, show that the minimum point of $f(x)$ is $\left(-\frac{1}{4}, \frac{7}{8}\right)$

To do this, a specific collection of tools available in Python will be used. Often specific sets of tools are separated into things called **libraries**. Start by telling Python that you want to use the specific library for **symbolic mathematics**:

DOI: 10.1201/9781003451860-3

```
Jupyter input
1    import sympy
```

This will allow you to solve the first part of the question. Create a variable `expression` and **assign** it a value of $\frac{1}{\sqrt{2}+1}$.

```
Jupyter input
1    expression = 1 / (sympy.sqrt(2) + 1)
2    expression
```

$$\frac{1}{1 + \sqrt{2}}$$

This is not what would happen if you plugged the above into a basic calculator, it would instead give you a value of:

```
Jupyter input
1    float(expression)
```

$$0.41421356237309503$$

The `sympy` library has a diverse set of tools available, one of which is to algorithmically attempt to simplify an expression. Here is how to do that:

```
Jupyter input
1    sympy.simplify(expression)
```

$$-1 + \sqrt{2}$$

This implies that:

$$\frac{1}{\sqrt{2} + 1} = -1 + \sqrt{2}$$

Multiplying both sides by $\sqrt{2} + 1$ gives:

$$1 = \frac{1}{\sqrt{2} + 1} \times \left(\sqrt{2} + 1\right) = \left(-1 + \sqrt{2}\right) \times \left(\sqrt{2} + 1\right)$$

The `sympy.simplify` command did not give much insight into what happened but you can confirm the above manipulation by expanding $\left(-1 + \sqrt{2}\right) \times \left(\sqrt{2} + 1\right)$. Here is how to do that:

```
Jupyter input
1    sympy.expand((-1 + sympy.sqrt(2)) * (1 + sympy.sqrt(2)))
```

$$1$$

The `sympy` library allows you to carry out basic expression manipulation. Now consider the second part of the question:

1. Consider the : $f(x) = 2x^2 + x + 1$:

2. Calculate the of the equation $2x^2 + x + 1 = 0$. What does this tell you about the solutions to the equation? What does this tell you about the graph of $f(x)$?

3. By completing the square, show that the minimum point of $f(x)$ is $\left(-\frac{1}{4}, \frac{7}{8}\right)$

Start by reassigning the value of the variable **expression** to be the expression: $2x^2 + x + 1$.

```
Jupyter input
1    x = sympy.Symbol("x")
2    expression = 2 * x ** 2 + x + 1
3    expression
```

$$2x^2 + x + 1$$

The first line communicates to the code that `x` is going to be a symbolic variable.

Recall that the `**` symbol is how you communicate exponentiation.

You can immediately use this to compute the:

```
Jupyter input
1    sympy.discriminant(expression)
```

$$-7$$

Now, complement this with mathematical knowledge: if a quadratic has a negative discriminant, then it does not have any roots and all the values are of the same sign as the coefficient of x^2. Which in this case is $2 > 0$. Confirm this by directly creating the equation. Do this by creating a variable **equation** and assigning it the equation which has a `lhs` and a `rhs`:

Jupyter input

```
1  equation = sympy.Eq(lhs=expression, rhs=0)
2  equation
```

$$2x^2 + x + 1 = 0$$

Now ask sympy to solve it:

Jupyter input

```
1  sympy.solveset(equation)
```

$$\left\{ -\frac{1}{4} - \frac{\sqrt{7}i}{4}, -\frac{1}{4} + \frac{\sqrt{7}i}{4} \right\}$$

Indeed the only solutions are imaginary numbers: this confirms that the graph of $f(x)$ is a convex parabola that is above the $y = 0$ line. Now complete the square so that you can write:

$$f(x) = a(x - b)^2 + c$$

for some values of a, b, c. Create variables that have those 3 constants as value but also create a variable `completed_square` and assign it the general expression:

Jupyter input

```
1  a, b, c = sympy.Symbol("a"), sympy.Symbol("b"), sympy.Symbol("c")
2  completed_square = a * (x - b) ** 2 + c
3  completed_square
```

$$a\left(-b + x\right)^2 + c$$

Expand this:

Jupyter input

```
1  sympy.expand(completed_square)
```

$$ab^2 - 2abx + ax^2 + c$$

Use sympy to solve the various equations that arise from comparing the coefficients of:

$$f(x) = 2x^2 + x + 1$$

with the completed square. First, you see that the coefficient of x^2 gives you an equation:

$$a = 2$$

For completeness write the code that solves this trivial equation:

Jupyter input

```
equation = sympy.Eq(a, 2)
sympy.solveset(equation, a)
```

$$\{2\}$$

Now substitute this value of a into the completed square and update the variable with the new value:

Jupyter input

```
completed_square = completed_square.subs({a: 2})
completed_square
```

$$c + 2\left(-b + x\right)^2$$

There are different types of brackets being used here: both () and {}. This is important and has specific meaning in Python which will be covered in future chapters.

Now look at the expression with the two remaining constants:

Jupyter input

```
sympy.expand(completed_square)
```

$$2b^2 - 4bx + c + 2x^2$$

Comparing the coefficients of x gives:

$$-4b = 1$$

```
1    equation = sympy.Eq(-4 * b, 1)
2    sympy.solveset(equation, b)
```

$$\left\{ -\frac{1}{4} \right\}$$

Substitute this value of b back into our expression. Make a point to tell sympy to treat $1/4$ symbolically and to not calculate the numeric value:

```
1    completed_square = completed_square.subs({b: -1 / sympy.S(4)})
2    completed_square
```

$$c + 2\left(x + \frac{1}{4} \right)^2$$

Expand this to see the expression with the one remaining constant gives:

```
1    sympy.expand(completed_square)
```

$$c + 2x^2 + x + \frac{1}{8}$$

This gives a final equation for the constant term:

$$c + 1/8 = 1$$

Now use sympy to find the value of c:

```
1    sympy.solveset(sympy.Eq(c + sympy.S(1) / 8, 1), c)
```

$$\left\{ \frac{7}{8} \right\}$$

As before substitute in and update the value of `completed_square`:

```
1   completed_square = completed_square.subs({c: 7 / sympy.S(8)})
2   completed_square
```

$$2\left(x + \frac{1}{4}\right)^2 + \frac{7}{8}$$

Using this shows that the there are indeed no values of x which give negative values of $f(x)$ as $f(x)$ is a square added to a constant. The minimum is when $x = -1/4$ which gives: $f(-1/4) = 7/8$:

```
1   completed_square.subs({x: -1 / sympy.S(4)})
```

$$\frac{7}{8}$$

This tutorial has:

- Created symbolic expressions.

- Obtained approximate values for numerical symbolic expressions.

- Expanded and simplified symbolic expressions.

- Created symbolic equations.

- Solve symbolic equations.

- Substituted values into symbolic expressions.

3.3   HOW TO

3.3.1   Create a symbolic numeric value

To create a symbolic numerical value use `sympy.S`.

```
1   sympy.S(a)
```

For example:

Jupyter input

```
1   import sympy
2
3   value = sympy.S(3)
4   value
```

$$3$$

If you combine a symbolic value with a non-symbolic value, it will automatically give a symbolic value:

Jupyter input

```
1   1 / value
```

$$\frac{1}{3}$$

3.3.2   Get the numerical value of a symbolic expression

You can get the numerical value of a symbolic value using `float` or `int`:

- `float` will give the numeric approximation in $\{R\}$

Usage

```
1   float(x)
```

- `int` will give the integer value

Usage

```
1   int(x)
```

For example, to create a symbolic numeric variable with value $frac15$:

Jupyter input

```
1   value = 1 / sympy.S(5)
2   value
```

$$\frac{1}{5}$$

To get the numerical value:

> **Jupyter input**
>
> 1 `float(value)`

`0.2`

To get the integer part:

> **Jupyter input**
>
> 1 `int(value)`

`0`

> This is not rounding to the nearest integer. It is returning the integer part.

3.3.3 Factor an expression

Use the `sympy.factor` tool to factor expressions.

> **Jupyter input**
>
> 1 `sympy.factor(expression)`

For example:

> **Jupyter input**
>
> 1 `x = sympy.Symbol("x")`
> 2 `sympy.factor(x ** 2 - 9)`

$$(x - 3)(x + 3)$$

3.3.4 Expand an expression

Use the `sympy.expand` tool to expand expressions.

> **Jupyter input**
>
> 1 `sympy.expand(expression)`

For example:

> **Jupyter input**
>
> 1 `sympy.expand((x - 3) * (x + 3))`

$$x^2 - 9$$

3.3.5 Simplify an expression

Use the `sympy.simplify` tool to simplify an expression.

> **Jupyter input**
>
> 1 `sympy.simplify(expression)`

For example:

> **Jupyter input**
>
> 1 `sympy.simplify((x - 3) * (x + 3))`

$$x^2 - 9$$

> This will not always give the expected (or any) result. At times it could be more beneficial to use `sympy.expand` and/or `sympy.factor`.

3.3.6 Solve an equation

Use the `sympy.solveset` tool to solve an equation. It takes two values as inputs. The first is either:

- An expression for which a root is to be found

- An equation

The second is the variable you want to solve for.

Usage

```
1   sympy.solveset(equation, variable)
```

Here is how you can use `sympy` to obtain the roots of the general:

$$ax^2 + bx + c$$

Jupyter input

```
1   a = sympy.Symbol("a")
2   b = sympy.Symbol("b")
3   c = sympy.Symbol("c")
4   quadratic = a * x ** 2 + b * x + c
5   sympy.solveset(quadratic, x)
```

$$\left\{ -\frac{b}{2a} - \frac{\sqrt{-4ac + b^2}}{2a}, -\frac{b}{2a} + \frac{\sqrt{-4ac + b^2}}{2a} \right\}$$

Here is to solve the same equation but not for x but for b:

Jupyter input

```
1   sympy.solveset(quadratic, b)
```

$$\left\{ -\frac{ax^2 + c}{x} \right\}$$

It is however clearer to specifically write the equation to solve:

Jupyter input

```
1   equation = sympy.Eq(a * x ** 2 + b * x + c, 0)
2   sympy.solveset(equation, x)
```

$$\left\{ -\frac{b}{2a} - \frac{\sqrt{-4ac + b^2}}{2a}, -\frac{b}{2a} + \frac{\sqrt{-4ac + b^2}}{2a} \right\}$$

3.3.7 Substitute a value into an expression

Given a `sympy` expression it is possible to substitute values into it using the `.subs()` tool.

> **Usage**
>
> ```
> 1  expression.subs({variable: value})
> ```

It is possible to pass multiple variables at a time. For example, to substitute the values for a, b, c into the :

> **Jupyter input**
>
> ```
> 1  quadratic = a * x ** 2 + b * x + c
> 2  quadratic.subs({a: 1, b: sympy.S(7) / 8, c: 0})
> ```

$$x^2 + \frac{7x}{8}$$

3.4 EXERCISES

If you are not sure how to do something, have a look at the "How To" section.

1. Simplify the following expressions.

    (a) $\frac{3}{\sqrt{3}}$

    (b) $\frac{2^{78}}{2^{12}2^{-32}}$

    (c) 8^0

    (d) $a^4 b^{-2} + a^3 b^2 + a^4 b^0$

2. Solve the following equations.

    (a) $x + 3 = -1$

    (b) $3x^2 - 2x = 5$

    (c) $x(x - 1)(x + 3) = 0$

    (d) $4x^3 + 7x - 24 = 1$

3. Consider the equation: $x^2 + 4 - y = \frac{1}{y}$.

    (a) Find the solution to this equation for x.

    (b) Obtain the specific solution when $y = 5$. Do this in two ways: substitute the value into your equation and substitute the value into your solution.

4. Consider: $f(x) = 4x^2 + 16x + 25$

    (a) Calculate the discriminant of the equation $4x^2 + 16x + 25 = 0$. What does this tell you about the solutions to the equation? What does this tell you about the graph of $f(x)$?

(b) By completing the square, show that the minimum point of $f(x)$ is $(-2, 9)$

5. Consider: $f(x) = -3x^2 + 24x - 97$

   (a) Calculate the discriminant of the equation $-3x^2 + 24x - 97 = 0$. What does this tell you about the solutions to the equation? What does this tell you about the graph of $f(x)$?

   (b) By completing the square, show that the maximum point of $f(x)$ is $(4, -49)$.

6. Consider the function $f(x) = x^2 + ax + b$.

   (a) Given that $f(0) = 0$ and $f(3) = 0$ obtain the values of a and b.

   (b) By completing the square, confirm that graph of $f(x)$ has a line of symmetry at $x = \frac{3}{2}$.

3.5   FURTHER INFORMATION

3.5.1   Why is some code in separate libraries?

When you run the `import sympy` command you are telling Python that you want to use a specific set of tools. You will see other examples of this throughout this book. One of the advantages of having code in libraries is that it is more efficient for Python to only use what is needed. There are two types of Python libraries:

- Those that are parts of the so-called "standard library": these are parts of Python itself.

- Those that are completely separate: `sympy` is one such example of this.

This separation allows for the development of tools to be independent of each other. The developers of `sympy` do not need to coordinate with the developers of Python to make new releases of the software.

3.5.2   Why do I need to use sympy?

`sympy` is the library for symbolic mathematics. There are other python libraries for carrying out mathematics in Python. For example, compute the value of the following expression:

$$(\sqrt{2} + 2)^2 - 2$$

You could compute this using the `math` library (for the square root tool):

```
1   import math
2
3   (math.sqrt(2) + 2) ** 2 - 2
```

9.65685424949238

You could also make use of the fact that you do not need a square root tool at all:

$$(\sqrt{2} + 2)^2 - 2 = (2^{1/2} + 2)^2 - 2$$

```
1   (2 ** (1 / 2) + 2) ** 2 - 2
```

$$9.65685424949238$$

You see that in both those instances, you have a numeric value for the expression that seems to be precise up to 14 decimal places.

However, that is **not** the exact value of that expression. The exact value of the expression needs to be computed symbolically:

```
1   import sympy
2
3   expression = (sympy.sqrt(2) + 2) ** 2 - 2
4   sympy.expand(expression)
```

$$4 + 4\sqrt{2}$$

This is one example of why **sympy** is an effective tool for mathematicians. The other one seen in this chapter is being able to compute expressions with no numerical value at all:

```
1   a = sympy.Symbol("a")
2   b = sympy.Symbol("b")
3   sympy.factor(a ** 2 - b ** 2)
```

$$(a - b)(a + b)$$

3.5.3 Why do I sometimes see `from sympy import *`?

There a number of resources available from which you can learn to use **sympy**. In some instances you will not see `import sympy` but instead you will see `from sympy import *`.

This it not a good way to do it.

What this does is taking all the tools inside of sympy and putting it at the same level of all the other tools available to you. The problem with doing this is that it no longer makes your code clear. An example of this are trigonometric functions. These exist in a number of libraries:

Jupyter input

```
1  import math
```

Jupyter input

```
1  import sympy
```

Jupyter input

```
1  sympy.cos(0)
```

$$1$$

Jupyter input

```
1  math.cos(0)
```

```
1.0
```

One of these tools allows you to carry out exact computations:

Jupyter input

```
1  sympy.cos(sympy.pi / 4)
```

$$\frac{\sqrt{2}}{2}$$

Jupyter input

```
1  math.cos(math.pi / 4)
```

```
0.7071067811865476
```

If you chose to import all the functionality using `from sympy import *`, then you cannot tell immediately which function you are using (except from its output):

```
Jupyter input

1  from sympy import *
```

```
Jupyter input

1  from math import *
```

```
Jupyter input

1  cos(pi / 4)
```

`0.7071067811865476`

In that case the second import has overwritten the first.

> **It is never recommended to use** `import *` which makes your code less clear and you are more likely to make mistakes when your code is not clear.

3.5.4   How to extract a solution from the output of `sympy.solveset`?

In some cases you might want to directly access the items in a solution set. For example, consider the equation $(x - 1)(x - 2)$.

```
Jupyter input

1  import sympy
2
3  x = sympy.Symbol("x")
4  expression = (x - 1) * (x - 2)
5  equation = sympy.Eq(expression, 0)
6  set_of_solutions = sympy.solveset(equation, x)
7  set_of_solutions
```

$$\{1, 2\}$$

The `set_of_solutions` has value the **set** of solutions of the equation. If you wanted to access them directly, you can use the following:

```
1   tuple_of_solutions = set_of_solutions.args
2   tuple_of_solutions
```

$$(1, 2)$$

This creates a **finite** ordered tuple of the solutions. You can use concepts that are covered in Chapter 6 and access them directly. Because there are two roots you can use the following to create two new variables:

```
1   x1, x2 = tuple_of_solutions
```

Substitute these values directly into the expression:

```
1   expression.subs({x: x1})
```

$$0$$

```
1   expression.subs({x: x2})
```

$$0$$

Note that this is not always possible to get a finite ordered tuple of the solutions, for example, there are some equations where the set of solutions is an infinite set:

```
1   equation = sympy.Eq(sym.cos(x / 5), 0)
2   set_of_solutions = sympy.solveset(equation, x)
3   set_of_solutions
```

$$\left\{ 10n\pi + \frac{5\pi}{2} \ \middle|\ n \in \mathbb{Z} \right\} \cup \left\{ 10n\pi + \frac{15\pi}{2} \ \middle|\ n \in \mathbb{Z} \right\}$$

3.5.5 Why do I sometimes see `import sympy as sym`?

In some resources you will see that instead of `import sympy` people use: `import sympy as sym`. This is called **aliasing**. This is common and takes advantage of the fact that Python can import a library and give it an alias/nickname at the same time:

Usage

```
1   import <library> as <nickname>
```

So with sympy you can use:

Jupyter input

```
1   import sympy as sym
2
3   sym.cos(sym.pi / 4)
```

$$\frac{\sqrt{2}}{2}$$

There is nothing stopping you using whatever alias you want:

Jupyter input

```
1   import sympy as a_poor_name_choice
2
3   a_poor_name_choice.cos(a_poor_name_choice.pi / 4)
```

$$\frac{\sqrt{2}}{2}$$

> **It is important** when aliasing to use accepted conventions for these nicknames. For sympy, an accepted convention is indeed `import sympy as sym`.

CHAPTER 4

Calculus

The A-level syllabus describes Calculus as:

> "Calculus: this is a fundamental element which describes change in dynamic situations and underlines the links between functions and graphs."

In practice this often means:

- taking of functions;
- functions;
- functions.

Here you will see how to instruct a computer to carry out these techniques.

In this chapter you will cover:

- Taking limits of functions.
- Differentiating functions.
- Computing definite and indefinite integrals.

4.1 TUTORIAL

You will solve the following problem using a computer to assist with the technical aspects:
Consider the function $f(x) = \frac{24x(a-4x)+2(a-8x)(b-4x)}{(b-4x)^4}$

1. Given that $\frac{df}{dx}\big|_{x=0} = 0$, $\frac{d^2f}{dx^2}\big|_{x=0} = -1$ and that $b > 0$ find the values of a and b.

2. For the specific values of a and b, find

   (a) $\lim_{x \to 0} f(x)$;
   (b) $\lim_{x \to \infty} f(x)$;
   (c) $\int f(x)dx$;
   (d) $\int_5^{20} f(x)dx$.

Sympy is once again the library you will use for this. You will start by creating a variable `expression` that has the value of the expression of $f(x)$:

DOI: 10.1201/9781003451860-4

```
Jupyter input

1    import sympy as sym
2
3    x = sym.Symbol("x")
4    a = sym.Symbol("a")
5    b = sym.Symbol("b")
6    expression = (24 * x * (a - 4 * x) + 2 * (a - 8 * x) * (b - 4 * x)) /
     ↪    ((b - 4 * x) ** 4)
7    expression
```

$$\frac{24x\left(a-4x\right)+\left(2a-16x\right)\left(b-4x\right)}{\left(b-4x\right)^4}$$

You will use `sympy.diff` to calculate the derivative. This tool takes two inputs:

- the first is the expression you are differentiating. Essentially this is the numerator of $\frac{df}{dx}$.

- the second is the variable you are differentiating with respect to. This is the denominator of $\frac{df}{dx}$.

You have imported `import sympy as sym` so you are going to write `sym.diff`:

```
Jupyter input

1    derivative = sym.diff(expression, x)
2    derivative
```

$$\frac{16a-16b-64x}{\left(b-4x\right)^4}+\frac{16\cdot\left(24x\left(a-4x\right)+\left(2a-16x\right)\left(b-4x\right)\right)}{\left(b-4x\right)^5}$$

Factorise that to make it slightly clearer:

```
Jupyter input

1    sym.factor(derivative)
```

$$\frac{16\left(-3ab-12ax+b^2+16bx+16x^2\right)}{\left(-b+4x\right)^5}$$

You will now create the first equation, which is obtained by substituting $x=0$ into the value of the derivative and equating that to 0:

Jupyter input

```
1   first_equation = sym.Eq(derivative.subs({x: 0}), 0)
2   first_equation
```

$$\frac{32a}{b^4} + \frac{16a - 16b}{b^4} = 0$$

Factor that equation:

Jupyter input

```
1   sym.factor(first_equation)
```

$$\frac{16 \cdot (3a - b)}{b^4} = 0$$

Now you are going to create the second equation, substituting $x = 0$ into the value of the second derivative. Calculate the second derivative by passing a third (optional) input to `sym.diff`:

Jupyter input

```
1   second_derivative = sym.diff(expression, x, 2)
2   second_derivative
```

$$\frac{64\left(-1 - \frac{8(-a+b+4x)}{b-4x} + \frac{10 \cdot (12x(a-4x)+(a-8x)(b-4x))}{(b-4x)^2}\right)}{(b - 4x)^4}$$

Equate this expression to -1:

Jupyter input

```
1   second_equation = sym.Eq(second_derivative.subs({x: 0}), -1)
2   second_equation
```

$$\frac{64 \cdot \left(\frac{10a}{b} - 1 - \frac{8(-a+b)}{b}\right)}{b^4} = -1$$

Now solve the first equation to obtain a value for a:

Jupyter input

```
1   sym.solveset(first_equation, a)
```

$$\left\{\frac{b}{3}\right\}$$

Now to substitute that value for a and solve the second equation for b:

Jupyter input

```
1   second_equation = second_equation.subs({a: b / 3})
2   second_equation
```

$$-\frac{192}{b^4} = -1$$

Jupyter input

```
1   sym.solveset(second_equation, b)
```

$$\left\{-2\sqrt{2} \cdot \sqrt[4]{3}, 2\sqrt{2} \cdot \sqrt[4]{3}, -2\sqrt{2} \cdot \sqrt[4]{3}i, 2\sqrt{2} \cdot \sqrt[4]{3}i\right\}$$

Recalling the question you know that $b > 0$, thus $b = 2\sqrt{2}\sqrt[4]{3}$ and $a = \frac{2\sqrt{2}\sqrt[4]{3}}{3}$. You will substitute these values back and finish the question:

Jupyter input

```
1   expression = expression.subs(
2       {
3           a: 2 * sym.sqrt(2) * sym.root(3, 4) / 3,
4           b: 2 * sym.sqrt(2) * sym.root(3, 4),
5       }
6   )
7   expression
```

$$\frac{24x\left(-4x + \frac{2\sqrt{2} \cdot \sqrt[4]{3}}{3}\right) + \left(-16x + \frac{4\sqrt{2} \cdot \sqrt[4]{3}}{3}\right)\left(-4x + 2\sqrt{2} \cdot \sqrt[4]{3}\right)}{\left(-4x + 2\sqrt{2} \cdot \sqrt[4]{3}\right)^4}$$

You are using the sym.root command for the generic nth root. You can confirm this:

Jupyter input

```
1  sym.diff(expression, x).subs({x: 0})
```

$$0$$

Jupyter input

```
1  sym.diff(expression, x, 2).subs({x: 0})
```

$$-1$$

Now you will calculate the using `sym.limit`, this takes three inputs:

- The expression you are taking the limit of.

- The variable that is changing.

- The value that the variable is tending towards.

Jupyter input

```
1  sym.limit(expression, x, 0)
```

$$\frac{\sqrt{3}}{36}$$

Jupyter input

```
1  sym.limit(expression, x, sym.oo)
```

$$0$$

Now you are going to calculate the **indefinite** integral using `sympy.integrate`. This tool takes two inputs as:

- the first is the expression you're integrating. This is the f in $\int_a^b f dx$.

- the second is the remaining information needed to calculate the integral: x.

```
1  sym.factor(sym.integrate(expression, x))
```

$$\frac{x\left(6x - \sqrt{2}\cdot\sqrt[4]{3}\right)}{12\cdot\left(4x^3 - 6\sqrt{2}\cdot\sqrt[4]{3}x^2 + 6\sqrt{3}x - \sqrt{2}\cdot 3^{\frac{3}{4}}\right)}$$

If you want to calculate a **definite** integral, then instead of passing the single variable you pass a tuple which contains the variables as the bounds of integration:

```
1  sym.factor(sym.integrate(expression, (x, 5, 20)))
```

$$-\frac{5\left(-5000\sqrt{2}\cdot\sqrt[4]{3} - 1200\sqrt{3} + 75\sqrt{2}\cdot 3^{\frac{3}{4}} + 119997\right)}{2\left(-32000 - 120\sqrt{3} + \sqrt{2}\cdot 3^{\frac{3}{4}} + 2400\sqrt{2}\cdot\sqrt[4]{3}\right)\left(-500 - 30\sqrt{3} + \sqrt{2}\cdot 3^{\frac{3}{4}} + 150\sqrt{2}\cdot\sqrt[4]{3}\right)}$$

This tutorial has:

- Simplified a rational quotient;

- Differentiated symbolic expressions;

- Solved algebraic equations.

4.2   HOW TO

4.2.1   Calculate the derivative of an expression.

You can calculate the derivative of an expression using `sympy.diff` which takes, an expression, a variable and a degree.

```
1  sympy.diff(expression, variable, degree=1)
```

The default value of **degree** is 1. For example, to compute $\frac{d(4x^3 + 2x + 1)}{dx}$:

```
1   import sympy as sym
2
3   x = sym.Symbol("x")
4   expression = 4 * x ** 3 + 2 * x + 1
5   sym.diff(expression, x)
```

$$12x^2 + 2$$

To compute the second derivative: $\frac{d^2(4x^3+2x+1)}{dx^2}$

Jupyter input

```
1   sym.diff(expression, x, 2)
```

$$24x$$

4.2.2 Calculate the indefinite integral of an expression.

You can calculate the indefinite integral of an expression using `sympy.integrate`, which takes an expression and a variable.

Usage

```
1   sympy.integrate(expression, variable)
```

For example, to compute $\int 4x^3 + 2x + 1\,dx$:

Jupyter input

```
1   sym.integrate(expression, x)
```

$$x^4 + x^2 + x$$

4.2.3 Calculate the definite integral of an expression.

You can calculate the definite integral of an expression using `sympy.integrate`. The first argument is an expression but instead of passing a variable as the second argument you pass a tuple with the variable as well as the upper and lower bounds of integration.

> **Usage**
>
> 1 `sympy.integrate(expression, (variable, lower_bound, upper_bound))`

For example, to compute $\int_0^4 4x^3 + 2x + 1 \, dx$:

> **Jupyter input**
>
> 1 `sym.integrate(expression, (x, 0, 4))`

276

4.2.4 Use ∞

In sympy you can access ∞ using `sym.oo`:

> **Usage**
>
> 1 `sympy.oo`

For example:

> **Jupyter input**
>
> 1 `sym.oo`

∞

4.2.5 Calculate limits of an expression

You can calculate using `sympy.limit`. The first argument is the expression, then the variable and finally the expression the variable tends to.

> **Usage**
>
> 1 `sympy.limit(expression, variable, value)`

For example, to compute $\lim_{h \to 0} \frac{4x^3 + 2x + 1 - 4(x-h)^3 - 2(x-h) - 1}{h}$:

Jupyter input

```
1  h = sym.Symbol("h")
2  expression = (4 * x ** 3 + 2 * x + 1 - 4 * (x - h) ** 3 - 2 * (x - h) -
   ↪  1) / h
3  sym.limit(expression, h, 0)
```

$$12x^2 + 2$$

4.3 EXERCISES

1. For each of the following functions calculate $\frac{df}{dx}$, $\frac{d^2 f}{dx^2}$ and $\int f(x)dx$.

   (a) $f(x) = x$

   (b) $f(x) = x^{\frac{1}{3}}$

   (c) $f(x) = 2x(x - 3)(\sin(x) - 5)$

   (d) $f(x) = 3x^3 + 6\sqrt{x} + 3$

2. Consider the function $f(x) = 2x + 1$. By differentiating **from first principles** show that $f'(x) = 2$.

3. Consider the second derivative $f''(x) = 6x + 4$ of some cubic function $f(x)$.

   (a) Find $f'(x)$

   (b) You are given that $f(0) = 10$ and $f(1) = 13$, find $f(x)$.

   (c) Find all the stationary points of $f(x)$ and determine their nature.

4. Consider the function $f(x) = \frac{2}{3}x^3 + bx^2 + 2x + 3$, where b is some undetermined coefficient.

   (a) Find $f'(x)$ and $f''(x)$.

   (b) You are given that $f(x)$ has a stationary point at $x = 2$. Use this information to find b.

   (c) Find the coordinates of the other stationary point.

   (d) Determine the nature of all stationary points.

5. Consider the functions $f(x) = -x^2 + 4x + 4$ and $g(x) = 3x^2 - 2x - 2$.

   (a) Create a variable `turning_points` which has value the turning points of $f(x)$.

   (b) Create variable `intersection_points` which has value of the points where $f(x)$ and $g(x)$ intersect.

   (c) Using your answers to part b. calculate the area of the region between f and g. Assign this value to a variable `area_betyouen`.

4.4 FURTHER INFORMATION

4.4.1 How can you plot a function

It is possible to plot a function using `sympy` using the `sympy.plot` function:

Usage

```
1  sympy.plot(expression)
```

So for example, here is a plot of $f(x) = x^2 + 3x + 1$:

Jupyter input

```
1  import sympy as sym
2
3  x = sym.Symbol("x")
4  sym.plot(x ** 2 + 3 * x + 1)
```

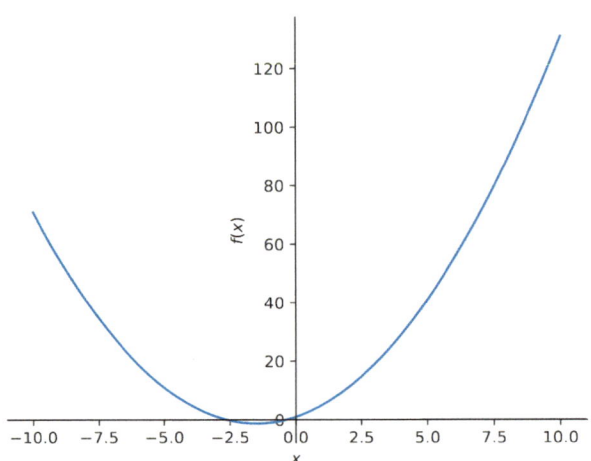

It is possible to specify the x and combine it with other plots:

Jupyter input

```
1  sym.plot(x ** 2 + 3 * x + 1, xlim=(-5, 5))
```

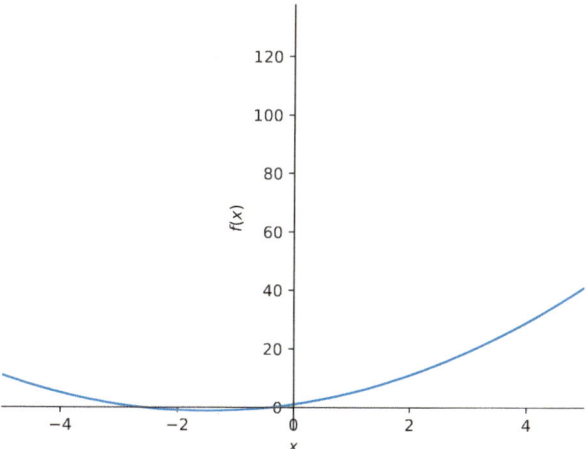

This plotting solution is good if you want to take a look at a function quickly but it is not recommended. The main python library for plotting is called `matplotlib` and is covered in a chapter of the online version of the book.

CHAPTER 5

Matrices

Matrices form the building block of an area of mathematics referred to as Linear Algebra. The dictionary definition of a matrix is:

> "A group of numbers or other symbols arranged in a rectangle that can be used together as a single unit to solve particular mathematical problems."

The particular mathematical problems referred to usually correspond to solving large systems of linear equations. However, they have become an area of interest in their own right and manipulating matrices usually corresponds to:

- calculating the determinant of a matrix

- calculating the inverse of a matrix

Here you will see how to instruct a computer to carry out these techniques. In this chapter you will cover:

- Creating matrices.

- Manipulating matrices.

- Solving a system of linear equations using matrices.

5.1   TUTORIAL

You will solve the following problem using a computer to assist with the technical aspects:

The matrix A is given by $A = \begin{pmatrix} a & 1 & 1 \\ 1 & a & 1 \\ 1 & 1 & 2 \end{pmatrix}$.

1. Find the determinant of A

2. Hence find the values of a for which A is singular.

3. For the following values of a, when possible obtain A^{-1} and confirm the result by computing AA^{-1}:

   (a) $a = 0$;

   (b) $a = 1$;

   (c) $a = 2$;

DOI: 10.1201/9781003451860-5

(d) $a = 3$.

sympy is once again the library you will use for this. You will start by defining the matrix A:

```
1  import sympy as sym
2
3  a = sym.symbol("a")
4  A = sym.matrix([[a, 1, 1], [1, a, 1], [1, 1, 2]])
```

You can now create a variable and assign it the value of the determinant of A:

```
1  determinant = A.det()
2  determinant
```

$$2a^2 - 2a$$

A matrix is singular if it has 0. You can find the values of a for which this occurs:

```
1  sym.solveset(determinant, a)
```

$$\{0, 1\}$$

Thus, it is not possible to find the inverse of A for $a \in \{0, 1\}$. However for $a = 2$:

```
1  A.subs({a: 2})
```

$$\begin{bmatrix} 2 & 1 & 1 \\ 1 & 2 & 1 \\ 1 & 1 & 2 \end{bmatrix}$$

```
1   A.subs({a: 2}).inv()
```

$$\begin{bmatrix} \frac{3}{4} & -\frac{1}{4} & -\frac{1}{4} \\ -\frac{1}{4} & \frac{3}{4} & -\frac{1}{4} \\ -\frac{1}{4} & -\frac{1}{4} & \frac{3}{4} \end{bmatrix}$$

To carry out matrix multiplication you use the @ symbol:

```
1   A.subs({a: 2}).inv() @ A.subs({a: 2})
```

$$\begin{bmatrix} 1 & 0 & 0 \\ 0 & 1 & 0 \\ 0 & 0 & 1 \end{bmatrix}$$

and for $a = 3$:

```
1   A.subs({a: 3}).inv()
```

$$\begin{bmatrix} \frac{5}{12} & -\frac{1}{12} & -\frac{1}{6} \\ -\frac{1}{12} & \frac{5}{12} & -\frac{1}{6} \\ -\frac{1}{6} & -\frac{1}{6} & \frac{2}{3} \end{bmatrix}$$

```
1   A.subs({a: 3}).inv() @ A.subs({a: 3})
```

$$\begin{bmatrix} 1 & 0 & 0 \\ 0 & 1 & 0 \\ 0 & 0 & 1 \end{bmatrix}$$

In this tutorial you have

- Created a matrix;

- Calculated the determinant of the matrix;

- Substituted values in the matrix; and

- Inverted the matrix.

5.2   HOW TO

5.2.1   Create a matrix

You create a matrix using the `sympy.Matrix` tool. Combine this with nested square brackets `[]` so that every row is also inside square brackets.

Usage

```
1    sympy.Matrix([values])
```

For example, the following creates the matrix:

$$B = \begin{pmatrix} 1 & 2 & 3 & 4 \\ 5 & 6 & 7 & 8 \\ 9 & 10 & 11 & 12 \end{pmatrix}$$

Jupyter input

```
1    import sympy as sym
2
3    B = sym.Matrix([[1, 2, 3, 4], [5, 6, 7, 8], [9, 10, 11, 12]])
4    B
```

$$\begin{bmatrix} 1 & 2 & 3 & 4 \\ 5 & 6 & 7 & 8 \\ 9 & 10 & 11 & 12 \end{bmatrix}$$

It is possible to write the code in a more readable way as long as an incomplete line ends with an open bracket:

Jupyter input

```
1   B = sym.Matrix(
2       [
3           [1, 2, 3, 4],
4           [5, 6, 7, 8],
5           [9, 10, 11, 12],
6       ]
7   )
```

5.2.2   Calculate the determinant of a matrix

To calculate the determinant of a matrix, use the `.det` tool.

Usage

```
1   matrix = sympy.Matrix([values])
2   matrix.det()
```

For example, the determinant of the following matrix:

$$\begin{pmatrix} 1 & 5 \\ 5 & 1 \end{pmatrix}$$

Jupyter input

```
1   matrix = sym.Matrix([[1, 5], [5, 1]])
2   matrix.det()
```

$$-24$$

5.2.3   Calculate the inverse of a matrix

To calculate the inverse of a matrix, use the `.inv` tool.

Usage

```
1   matrix = sympy.Matrix([values])
2   matrix.inv()
```

For example, to calculate the inverse of:

$$\begin{pmatrix} 1/2 & 1 \\ 5 & 0 \end{pmatrix}$$

```
1   matrix = sym.Matrix([[sym.S(1) / 2, 1], [5, 0]])
2   matrix.inv()
```

$$\begin{bmatrix} 0 & \frac{1}{5} \\ 1 & -\frac{1}{10} \end{bmatrix}$$

5.2.4 Multiply matrices by a scalar

To multiple a matrix by a scalar use the * operator. For example, to multiply the following matrix by 6:

$$\begin{pmatrix} 1/5 & 1 \\ 1 & 1 \end{pmatrix}$$

```
1   matrix = sym.Matrix([[sym.S(1) / 5, 1], [1, 1]])
2   6 * matrix
```

$$\begin{bmatrix} \frac{6}{5} & 6 \\ 6 & 6 \end{bmatrix}$$

5.2.5 Add matrices together

To add matrices use the + operator. For example to compute:

$$\begin{pmatrix} 1/5 & 1 \\ 1 & 1 \end{pmatrix} + \begin{pmatrix} 4/5 & 0 \\ 0 & 0 \end{pmatrix}$$

```
1   matrix = sym.Matrix([[sym.S(1) / 5, 1], [1, 1]])
2   other_matrix = sym.Matrix([[sym.S(4) / 5, 0], [0, 0]])
3   matrix + other_matrix
```

$$\begin{bmatrix} 1 & 1 \\ 1 & 1 \end{bmatrix}$$

5.2.6 Multiply matrices together

To multiply matrices together you use the @ operator. For example, to compute:

$$\begin{pmatrix} 1/5 & 1 \\ 1 & 1 \end{pmatrix} \begin{pmatrix} 4/5 & 0 \\ 0 & 0 \end{pmatrix}$$

Jupyter input

```
1   matrix @ other_matrix
```

$$\begin{bmatrix} \frac{4}{25} & 0 \\ \frac{4}{5} & 0 \end{bmatrix}$$

5.2.7 Create a vector

A vector is essentially a matrix with a single row or column. For example, to create the vector:

$$\begin{pmatrix} 3 \\ 2 \\ 1 \end{pmatrix}$$

Jupyter input

```
1   vector = sym.Matrix([[3], [2], [1]])
2   vector
```

$$\begin{bmatrix} 3 \\ 2 \\ 1 \end{bmatrix}$$

5.2.8 Solve a linear system

To solve a given linear system that can be represented in matrix form, create the corresponding matrix and vector and the matrix. For example, to solve the following equations:

$$x + 2y = 3$$
$$3x + y + 2z = 4$$
$$-y + z = 1$$

Jupyter input

```
1   A = sym.Matrix([[1, 2, 0], [3, 1, 2], [0, -1, 1]])
2   b = sym.Matrix([[3], [4], [1]])
3   A.inv() @ b
```

$$\begin{bmatrix} -\frac{5}{3} \\ \frac{7}{3} \\ \frac{10}{3} \end{bmatrix}$$

5.3 EXERCISES

1. Obtain the determinant and the inverse of the following matrices:

   (a) $A = \begin{pmatrix} 1/5 & 1 \\ 1 & 1 \end{pmatrix}$

   (b) $B = \begin{pmatrix} 1/5 & 1 & 5 \\ 3 & 1 & 6 \\ 1 & 2 & 1 \end{pmatrix}$

   (c) $C = \begin{pmatrix} 1/5 & 5 & 5 \\ 3 & 1 & 7 \\ a & b & c \end{pmatrix}$

2. Compute the following:

   (a) $500 \begin{pmatrix} 1/5 & 1 \\ 1 & 1 \end{pmatrix}$

   (b) $\pi \begin{pmatrix} 1/\pi & 2\pi \\ 3/\pi & 1 \end{pmatrix}$

   (c) $500 \begin{pmatrix} 1/5 & 1 \\ 1 & 1 \end{pmatrix} + \pi \begin{pmatrix} 1/\pi & 2\pi \\ 3/\pi & 1 \end{pmatrix}$

   (d) $500 \begin{pmatrix} 1/5 & 1 \\ 1 & 1 \end{pmatrix} \begin{pmatrix} 1/\pi & 2\pi \\ 3/\pi & 1 \end{pmatrix}$

3. The matrix A is given by $A = \begin{pmatrix} a & 4 & 2 \\ 1 & a & 0 \\ 1 & 2 & 1 \end{pmatrix}$.

   (a) Find the determinant of A

   (b) Hence find the values of a for which A is singular.

   (c) State, giving a brief reason in each case, whether the simultaneous equations

   $$\begin{aligned} ax + 4y + 2z &= 3a \\ x + ay &= 1 \\ x + 2y + z &= 3 \end{aligned}$$

   have any solutions when:

      i. $a = 3$

     ii. $a = 2$

4. The matrix D is given by $D = \begin{pmatrix} a & 2 & 0 \\ 3 & 1 & 2 \\ 0 & -1 & 1 \end{pmatrix}$ where $a \neq 2$.

  (a) Find D^{-1}.

  (b) Hence or otherwise, solve the equations:

$$ax + 2y = 3$$
$$3x + y + 2z = 4$$
$$-y + z = 1$$

5.4 FURTHER INFORMATION

5.4.1 Why does this book not discuss commenting of code?

In Python it is possible to write statements that are ignored using the # symbol. This creates something called a "comment". For example:

```
Jupyter input

1   import sympy as sym  # Importing the sympy library using an alias
```

Comments like these often do not add to the readability of the code. In fact they can make the code less readable or at worse confusing [4].

In this section of the book there is in fact no need for comments like this as you are mainly using tools that are well-documented. Furthermore when using Jupyter notebooks you can add far more to the readability of the code by adding prose alongside our code instead of using small brief inline comments.

This does not mean that readability of code is not important.

> Being able to read and understand written code is important.

In Chapter 12, you will start to write functions and emphasis will be given there on readability and documenting (as opposed to commenting) the code written. A specific discussion about using a tool called a **docstring** as opposed to a comment will be covered.

In Chapters 15 and 17, there is more information on how to ensure code is readable and understandable.

5.4.2 Why use @ for matrix multiplication and not *?

With sympy it is in fact possible to use the * operator for matrix multiplication:

Jupyter input

```
1   import sympy as sym
2
3   matrix = sym.Matrix([[sym.S(1) / 5, 1], [1, 1]])
4   other_matrix = sym.Matrix([[sym.S(4) / 5, 0], [0, 0]])
5   matrix * other_matrix
```

$$\begin{bmatrix} \frac{4}{25} & 0 \\ \frac{4}{5} & 0 \end{bmatrix}$$

However, there are other libraries that can be used for linear algebra and in those libraries the * does not do matrix multiplication, it does element-wise multiplication instead. So for clarity it is preferred to use @ throughout.

5.4.3 Is Numpy a library that can be used for linear algebra?

numpy is one of the most popular and important libraries in the Python ecosystem. It is in fact the best library to use when doing linear algebra as it is computationally efficient, **however**, it cannot handle symbolic variables which is why you are seeing how to use Sympy here. An introduction to numpy is covered in a chapter of the online version of the book.

CHAPTER 6

Combinatorics

Combinatorics is the mathematical area interested in specific finite sets. In a lot of instances this comes down to counting things and is often first encountered by mathematicians through combinations and permutations. Computers are useful when doing this as they can be used to generate the finite sets considered. You can essentially count things "by hand" (using a computer) to confirm theoretic results.

> In this chapter you will cover:
>
> - Generating configurations of elements that correspond to permutations and/or combinations.
>
> - Counting these configurations.
>
> - Directly computing $^nC_i = \binom{n}{i}$.
>
> - Directly computing ^nP_i.

6.1 TUTORIAL

You will solve the following problem using a computer to illustrate how a computer can be used to solve combinatorial problems:

The digits 1, 2, 3, 4, and 5 are arranged in random order, to form a five-digit number.

1. How many different five-digit numbers can be formed?

2. How many different five-digit numbers are:

   (a) Odd
   (b) Less than 23000

Firstly you are going to get the 5 digits. Python has a tool for this called `range` which directly gives the integers from a given bound to another:

```
Jupyter input
1  digits = range(1, 6)
2  digits
```

DOI: 10.1201/9781003451860-6

```
range(1, 6)
```

At present that is only the instructions for generating the integers from 1 to 5 (the 6 is a strict upper bound). You can transform this to a tuple, using the `tuple` tool:

> **Jupyter input**
>
> ```
> 1   tuple(range(1, 6))
> ```

```
(1, 2, 3, 4, 5)
```

The question is asking for all the permutations of size 5 of that set. The main tool for this is a library specifically designed to iterate over objects in different ways: `itertools`.

> **Jupyter input**
>
> ```
> 1   import itertools
> 2
> 3   permutations = itertools.permutations(digits)
> 4   permutations
> ```

```
<itertools.permutations at 0x103a548b0>
```

That is again only the set of instructions, to view the actual permutations you will transform this into a tuple. You will overwrite the value of `permutations` to not be the instructions but the actual tuple of all the permutations:

> **Jupyter input**
>
> ```
> 1   permutations = tuple(permutations)
> 2   permutations
> ```

```
((1, 2, 3, 4, 5),
 (1, 2, 3, 5, 4),
 (1, 2, 4, 3, 5),
 (1, 2, 4, 5, 3),
 ...
 (5, 4, 2, 3, 1),
 (5, 4, 3, 1, 2),
 (5, 4, 3, 2, 1))
```

Now to answer the question you need to find out how many tuples are in that tuple. You do this using the Python `len` tool which returns the length of something:

Jupyter input

```
1  len(permutations)
```

120

You can confirm this to be correct as you know that there are 5! ways of arranging those numbers. The math library has a `factorial` tool:

Jupyter input

```
1  import math
2
3  math.factorial(5)
```

120

In order to find out how many 5 digit numbers are odd you are going to compute the following sum:

$$\sum_{\pi \in \Pi} \pi_5 \mod 2$$

Where Π is the set of permutations and π_5 denotes the 5th (and last) element of the permutation. So for example, if the first element of Π was $(1, 2, 3, 4, 5)$ then $\pi_5 = 5$ and $5 \mod 2 = 1$. To do this, you use the `sum` tool in python coupled with the expressions `for` and `in`. You also access the 5th element of a given `permutation` using `[4]` (the first element is indexed by 0, so the 5th is indexed by 4):

Jupyter input

```
1  sum(permutation[4] % 2 for permutation in permutations)
```

72

You can again check this theoretically, there are three valid choices for the last digit of a given tuple to be odd: 1, 3 and 5. For each of those, there are then four choices for the remaining digits:

Jupyter input

```
1  math.factorial(4) * 3
```

72

To compute the number of digits that are less than or equal to 23000 you compute a similar sum except you use the <= operator and also convert the tuple of digits to a number in base 10:

$$(\pi_1, \pi_2, \pi_3, \pi_4, \pi_5) \to \pi_1 10^4 + \pi_2 10^3 + \pi_3 10^2 + \pi_4 10 + \pi_5$$

Thus you are going to compute the following sum:

$$\sum_{\pi \in \Pi \text{ if } \pi_1 10^4 + \pi_2 10^3 + \pi_3 10^2 + \pi_4 10 + \pi_5 \leq 23000} 1$$

Jupyter input

```
sum(
    1
    for permutation in permutations
    if permutation[0] * 10 ** 4
    + permutation[1] * 10 ** 3
    + permutation[2] * 10 ** 2
    + permutation[3] * 10
    + permutation[4]
    <= 23000
)
```

30

You can again confirm this theoretically, for a given tuple to be less than 23000 that is only possible if the first digit is 1 or 2:

- If it is 1, then any of the other 4! permutations of the other digits is valid;

- If it is 2, then the second digit must be 1 and any of the other 3! permutations of the other digits is valid.

Jupyter input

```
(math.factorial(4) + math.factorial(3))
```

30

In this tutorial you have

- Created permutations of a given tuples;

- Created permutations of a given tuples that obey a given condition;

- Counted how many permutations exist; and

- Directly computed the expected number of permutations.

6.2 HOW TO

6.2.1 Create a tuple

To create a tuple which is an ordered collection of objects that cannot be changed, use the () brackets.

> **Usage**
>
> 1 collection = (value_1, value_2, value_3, ..., value_n)

For example:

> **Jupyter input**
>
> 1 basket = ("Bread", "Biscuits", "Coffee")
> 2 basket

('Bread', 'Biscuits', 'Coffee')

6.2.2 How to access particular elements in a tuple

If you need to you can access elements of a collection using [] brackets. The first element has index 0:

> **Usage**
>
> 1 tuple[index]

For example:

> **Jupyter input**
>
> 1 basket[1]

'Biscuits'

6.2.3 Create boolean variables

A boolean variable has one of two values: True or False.
  To create a boolean variable here are some of the things you can use:

- Equality: value == other_value

- Inequality value != other_value

- Strictly less than value < other_value

- Less than or equal `value <= other_value`

- Inclusion `value in iterable`

This a subset of the operators available. For example:

```
Jupyter input
1  value = 5
2  other_value = 10
3
4  value == other_value
```

False

```
Jupyter input
1  value != other_value
```

True

```
Jupyter input
1  value <= other_value
```

True

```
Jupyter input
1  value < value
```

False

```
Jupyter input
1  value <= value
```

True

```
Jupyter input
1  value in (1, 2, 4, 19)
```

False

It is also possible to combine booleans to create new booleans:

- And: first_boolean and second_boolean
- Or: first_boolean or second_boolean
- No: not boolean

Jupyter input

```
1   True and True
```

True

Jupyter input

```
1   False and True
```

False

Jupyter input

```
1   True or False
```

True

Jupyter input

```
1   False or False
```

False

Jupyter input

```
1   not True
```

False

Jupyter input

```
1  not False
```

True

6.2.4 Create an iterable with a given number of items

The range tool gives a given number of integers.

Usage

```
1  range(number_of_integers)
```

For example:

Jupyter input

```
1  tuple(range(10))
```

(0, 1, 2, 3, 4, 5, 6, 7, 8, 9)

range(N) gives the integers from 0 until $N - 1$ (inclusive).
It is also possible to pass two values as inputs so that you have a different lower bound:

Usage

```
1  tuple(range(4, 10))
```

(4, 5, 6, 7, 8, 9)

It is also possible to pass a third value as a step size:

Jupyter input

```
1  tuple(range(4, 10, 3))
```

(4, 7)

6.2.5 Create permutations of a given set of elements

The python itertools library has a permutations tool that will generate all permutations of a given set.

> **Usage**
>
> 1 `itertools.permutations(iterable)`

> **Jupyter input**
>
> ```
> 1  import itertools
> 2
> 3  basket = ("Bread", "Biscuits", "Coffee")
> 4  tuple(itertools.permutations(basket))
> ```

```
(('Bread', 'Biscuits', 'Coffee'),
 ('Bread', 'Coffee', 'Biscuits'),
 ('Biscuits', 'Bread', 'Coffee'),
 ('Biscuits', 'Coffee', 'Bread'),
 ('Coffee', 'Bread', 'Biscuits'),
 ('Coffee', 'Biscuits', 'Bread'))
```

It is possible to limit the size to only be permutations of size r:

> **Jupyter input**
>
> ```
> 1  tuple(itertools.permutations(basket, r=2))
> ```

```
(('Bread', 'Biscuits'),
 ('Bread', 'Coffee'),
 ('Biscuits', 'Bread'),
 ('Biscuits', 'Coffee'),
 ('Coffee', 'Bread'),
 ('Coffee', 'Biscuits'))
```

6.2.6 Create combinations of a given set of elements

The python `itertools` library has a `combinations` tool that will generate all combinations of size r of a given set:

> **Usage**
>
> 1 `itertools.combinations(iterable, r)`

For example:

> **Jupyter input**
>
> ```
> 1  basket = ("Bread", "Biscuits", "Coffee")
> 2  tuple(itertools.combinations(basket, r=2))
> ```

(('Bread', 'Biscuits'), ('Bread', 'Coffee'), ('Biscuits', 'Coffee'))

A combination does not care about order so by default the combinations generated are sorted.

6.2.7   Summing items in an iterable

You can compute the sum of items in an iterable using the sum tool:

> **Jupyter input**
>
> ```
> 1  sum((1, 2, 3))
> ```

6

You can also directly use the sum without specifically creating the iterable. This corresponds to the following mathematical notation:

$$\sum_{s \in S} f(s)$$

and is done using the following:

> **Jupyter input**
>
> ```
> 1  sum(f(object) for object in old_list)
> ```

Here is an example of calculating the following sum:

$$\sum_{n=0}^{10} n^2$$

> **Jupyter input**
>
> ```
> 1  sum(n ** 2 for n in range(11))
> ```

385

You can compute conditional sums by only summing over elements that meet a given condition using the following:

> ### Usage
> ```
> 1    sum(f(object) for object in old_list if condition)
> ```

Here is an example of calculating the following sum:

$$\sum_{n=0}^{10} n^2$$
$$\text{if } n \text{ odd}$$

> ### Jupyter input
> ```
> 1    sum(n ** 2 for n in range(11) if n % 2 == 1)
> ```

165

6.2.8   Directly compute $n!$

The math library has a `factorial` tool.

> ### Usage
> ```
> 1    math.factorial(N)
> ```

> ### Jupyter input
> ```
> 1    import math
> 2
> 3    math.factorial(5)
> ```

120

6.2.9   Directly compute $\binom{n}{i}$

The `scipy.special` library has a `comb` tool.

> ### Usage
> ```
> 1    scipy.special.comb(n, i)
> ```

For example:

> **Jupyter input**
>
> ```
> 1  import scipy.special
> 2
> 3  scipy.special.comb(3, 2)
> ```

3.0

6.2.10 Directly compute ^nP_i

The `scipy.special` library has a `perm` tool.

> **Usage**
>
> ```
> 1  scipy.special.perm(n, i)
> ```

For example:

> **Jupyter input**
>
> ```
> 1  scipy.special.perm(3, 2)
> ```

6.0

6.3 EXERCISES

1. Obtain the following tuples using the `range` command:

   (a) $(0, 1, 2, 3, 4, 5)$

   (b) $(2, 3, 4, 5)$

   (c) $(2, 4, 6, 8)$

   (d) $-1, 2, 5, 8$

2. By **both** generating and directly computing obtain the **number of** the following:

   (a) All permutations of $(0, 1, 2, 3, 4, 5)$.

   (b) All permutations of (A, B, C).

   (c) Permutations of size 3 of $(0, 1, 2, 3, 4, 5)$.

   (d) Permutations of size 2 of $(0, 1, 2, 3, 4, 5, 6)$.

   (e) Combinations of size 3 of $(0, 1, 2, 3, 4, 5)$.

   (f) Combinations of size 2 of $(0, 1, 2, 3, 4, 5)$.

   (g) Combinations of size 5 of $(0, 1, 2, 3, 4, 5)$.

3. A class consists of 3 students from Ashville and 4 from Bewton. A committee of 5 students is chosen at random from the class.

(a) Find the number of committees that include 2 students from Ashville and 3 from Bewton are chosen.

(b) In fact, 2 students from Ashville and 3 from Bewton are chosen. In order to watch a video, all 5 committee members sit in a row. In how many different orders can they sit if no 2 students from Bewton sit next to each other.

4. Three letters are selected at random from the 8 letters of the word COMPUTER, without regard to order.

(a) Find the number of possible selections of 3 letters.

(b) Find the number of selections of 3 letters with the letter P.

(c) Find the number of selections of 3 letters where the 3 letters form the word TOP.

6.4   FURTHER INFORMATION

6.4.1   Are there other ways to access elements in tuples?

You have seen in this chapter how to access a single element in a tuple. There are various ways of indexing tuples:

1. Indexing (seen in Section 6.2.2).

2. Negative indexing (see Section 12.2.13)

3. Slicing (see Section 12.2.14)

6.4.2   Why do range, `itertools.permutations`, and `itertools.combinations` not directly give the elements?

When you run either of the three `range`, `itertools.permutations` or `itertools.combinations` tools, this is an example of creating a **generator**. This allows the creation of the instructions to build something without building it.

In practice this means that you can create large sets without needing to generate them until required.

6.4.3   How does the summation notation $\sum$ correspond to the code?

The `sum` command corresponds to the mathematical $\sum$ notation. Here are a few examples showing the `sum` command, the $\sum$ notation but also the prose describing:

- **The sum of the square of the integers from 1 to 100 (inclusive):**

$$\sum_{i=1}^{100} i^2$$

Given by:

```
Jupyter input

1   sum(i ** 2 for i in range(1, 101))
```

338350

- The sum of the square of the integers from 1 to 100 (inclusive) if they are prime:

$$\sum_{\substack{i=1}}^{100} i^2$$
if i is prime

Given by:

> **Jupyter input**
>
> ```
> 1   sum(i ** 2 for i in range(1, 101) if sym.isprime(i))
> ```

65796

- The sum of the square of the elements in the collection S if they are prime:

$$\sum_{\substack{i \in S}} i^2$$
if i is prime

Given by:

> **Jupyter input**
>
> ```
> 1   S = (1, 3, 9, 12, 21, 5, 2, 2)
> 2   sum(i ** 2 for i in S if sym.isprime(i))
> ```

42

CHAPTER 7

Probability

Probability is the study of random events. Computers are particularly helpful here as they can be used to carry out a number of experiments to confirm and/or explore theoretic results.

In practice studying probability will often involve measuring:

- expected chances of an event occurring and

- the conditional chances of an event occurring given another event occurring.

Here you will see how to instruct a computer to sample such events.

In this chapter you will cover:

- Generating random numbers.

- Randomly sampling from a given collection of items.

- Write python functions to be able to repeat experiments.

7.1 TUTORIAL

You will solve the following problem using a computer to estimate the expected probabilities:

An experiment consists of selecting a token from a bag and spinning a coin. The bag contains 5 red tokens and 7 blue tokens. A token is selected at random from the bag, its colour is noted and then the token is returned to the bag.

When a red token is selected, a biased coin with probability $\frac{2}{3}$ of landing heads is spun. When a blue token is selected a fair coin is spun.

1. What is the probability of picking a red token?

2. What is the probability of obtaining Heads?

3. If a heads is obtained, what is the probability of having selected a red token.

You will use the `random` library from the Python standard library to do this. First start off by building a Python **tuple** to represent the bag with the tokens. Assign this to a variable bag:

DOI: 10.1201/9781003451860-7

Jupyter input

```
1   bag = (
2        "Red",
3        "Red",
4        "Red",
5        "Red",
6        "Red",
7        "Blue",
8        "Blue",
9        "Blue",
10       "Blue",
11       "Blue",
12       "Blue",
13       "Blue",
14   )
15   bag
```

```
('Red',
 'Red',
 'Red',
 'Red',
 'Red',
 'Blue',
 'Blue',
 'Blue',
 'Blue',
 'Blue',
 'Blue',
 'Blue')
```

You are using the circular brackets () and the quotation marks ". Those are important and cannot be omitted. The choice of brackets () as opposed to {} or [] is important as it instructs Python to do different things. You can use " or ' interchangeably.

Instead of writing every copy of colour you can create a Python **list** which allows you to carry out some basic algebra on the items:

- Create a list with 5 "Red"s.

- Create a list with 7 "Blue"s.

- Combine both lists:

Jupyter input

```
1   bag = ["Red"] * 5 + ["Blue"] * 7
2   bag
```

```
['Red',
 'Red',
 'Red',
 'Red',
 'Red',
 'Blue',
 'Blue',
 'Blue',
 'Blue',
 'Blue',
 'Blue',
 'Blue']
```

Now to sample from that use the `random` library which has a `choice` command:

Jupyter input

```
1  import random
2
3  random.choice(bag)
```

```
'Blue'
```

If you run this many times, you will not always get the same outcome:

Jupyter input

```
1  random.choice(bag)
```

Jupyter input

```
1  'Blue'
```

The `bag` variable is unchanged:

Jupyter input

```
1  bag
```

Jupyter input

```
1  ['Red',
2   'Red',
3   'Red',
4   'Red',
5   'Red',
6   'Blue',
7   'Blue',
8   'Blue',
9   'Blue',
10  'Blue',
11  'Blue',
12  'Blue']
```

In order to answer the first question (what is the probability of picking a red token) repeat this many times: Do this by defining a Python function (which is akin to a mathematical function) that makes repeating code possible:

Jupyter input

```
1  def pick_a_token(container):
2      """
3      A function to randomly sample from a `container`.
4      """
5      return random.choice(container)
```

You can then call this function, passing `bag` to it as the `container` from which to pick:

Jupyter input

```
1  pick_a_token(container=bag)
```

```
'Blue'
```

Jupyter input

```
1  pick_a_token(container=bag)
```

```
'Red'
```

In order to measure the probability of picking a red token repeat this not once or twice but tens of thousands of times. You will do this using something called a "list comprehension" which is akin to the mathematical notation commonly used to create sets:

$$S_1 = \{f(x) \text{ for } x \text{ in } S_2\}$$

Jupyter input

```
1  number_of_repetitions = 10000
2  samples = [pick_a_token(container=bag) for repetition in
   ↪  range(number_of_repetitions)]
3  samples
```

Jupyter input

```
1  ['Red',
2   'Red',
3   'Red',
4   ...
5   'Blue',
6   'Blue',
7   'Red',
8   'Blue',
9  ]
```

You can confirm that you have the correct number of samples:

Jupyter input

```
1  len(samples)
```

```
10000
```

> `len` is the Python tool to get the length of a given Python iterable.

Using this you can now use `==` (double `=`) to check how many of those samples are Red:

Jupyter input

```
1  sum(token == "Red" for token in samples) / number_of_repetitions
```

```
0.4071
```

You have sampled a probability of around .41. The theoretic value is $\frac{5}{5+7}$:

Jupyter input

```
1   5 / (5 + 7)
```

0.4166666666666667

To answer the second question (What is the probability of obtaining Heads?). You need to make use of another Python tool: an `if` statement. This will let you write a function that does precisely what is described in the problem:

- Choose a token;

- Set the probability of flipping a given coin; and

- Select that coin.

For the second random selection (flipping a coin) you will not choose from a list but instead select a random number between 0 and 1.

Jupyter input

```
1   def sample_experiment(bag):
2       """
3       This samples a token from a given bag and then
4       selects a coin with a given probability.
5
6       If the sampled token is red then the probability
7       of selecting heads is 2/3 otherwise it is 1/2.
8
9       This function returns both the selected token
10      and the coin face.
11      """
12      selected_token = pick_a_token(container=bag)
13
14      if selected_token == "Red":
15          probability_of_selecting_heads = 2 / 3
16      else:
17          probability_of_selecting_heads = 1 / 2
18
19      if random.random() < probability_of_selecting_heads:
20          coin = "Heads"
21      else:
22          coin = "Tails"
23      return selected_token, coin
```

Using this you can sample according to the problem description:

Jupyter input

```
1   sample_experiment(bag=bag)
```

('Red', 'Heads')

Jupyter input

```
1   sample_experiment(bag=bag)
```

('Red', 'Tails')

You can now find out the probability of selecting heads by carrying out a large number of repetitions and checking which ones have a coin that is heads:

Jupyter input

```
1   samples = [sample_experiment(bag=bag) for repetition in
    ↪  range(number_of_repetitions)]
2   sum(coin == "Heads" for token, coin in samples) / number_of_repetitions
```

0.576

You can compute this theoretically as well, the expected probability is:

Jupyter input

```
1   import sympy as sym
2
3   sym.S(5) / (12) * sym.S(2) / 3 + sym.S(7) / (12) * sym.S(1) / 2
```

$$\frac{41}{72}$$

Jupyter input

```
1   41 / 72
```

0.5694444444444444

You can also use the samples to calculate the conditional probability that a token was read if the coin is heads. This is done again using the list comprehension notation but including an `if` statement which emulates the mathematical notation:

$$S_3 = \{x \in S_1 | \text{ if some property of } x \text{ holds}\}$$

```
1  samples_with_heads = [(token, coin) for token, coin in samples if coin
   ↪  == "Heads"]
2  sum(token == "Red" for token, coin in samples_with_heads) /
   ↪  len(samples_with_heads)
```

0.49236111111111114

This is given theoretically by:

$$P(\text{Red}|\text{Heads}) = \frac{P(\text{Heads}|\text{Red})P(\text{Red})}{P(\text{Heads})}$$

Jupyter input

```
1  (sym.S(2) / 3 * sym.S(5) / 12) / (sym.S(41) / 72)
```

$$\frac{20}{41}$$

Jupyter input

```
1  20 / 41
```

0.4878048780487805

In this tutorial you have

- Randomly sampled from an iterable.

- Randomly sampled a number between 0 and 1.

- Written a function to represent a random experiment.

- Created a list using list comprehensions.

- Counted outcomes of random experiments.

7.2 HOW TO

7.2.1 Create a list

To create a list which is an ordered collection of objects that **can** be changed use the [] brackets.

> **Usage**
>
> ```
> 1  collection = [value_1, value_2, value_3, ..., value_n]
> ```

For example:

> **Jupyter input**
>
> ```
> 1  basket = ["Bread", "Biscuits", "Coffee"]
> 2  basket
> ```

```
['Bread', 'Biscuits', 'Coffee']
```

You can insert an element to the end of a list by appending to it:

> **Jupyter input**
>
> ```
> 1  basket.append("Tea")
> 2  basket
> ```

```
['Bread', 'Biscuits', 'Coffee', 'Tea']
```

You can also combine lists together:

> **Jupyter input**
>
> ```
> 1  other_basket = ["Toothpaste"]
> 2  basket = basket + other_basket
> 3  basket
> ```

> **Jupyter input**
>
> ```
> 1  ['Bread', 'Biscuits', 'Coffee', 'Tea', 'Toothpaste']
> ```

As for tuples you can also access elements using their indices:

> **Jupyter input**
>
> ```
> 1  basket[3]
> ```

Jupyter input

```
1  'Tea'
```

7.2.2 Define a function

Define a function using the `def` keyword (short for define):

Usage

```
1  def name(variable1, variable2, ...):
2      """
3      A docstring between triple quotation to describe what is happening
4      """
5      INDENTED BLOCK OF CODE
6      return output
```

For example, define $f : \mathbb{R} \to \mathbb{R}$ given by $f(x) = x^3$ using the following:

Jupyter input

```
1  def x_cubed(x):
2      """
3      A function to return x ^ 3
4      """
5      return x ** 3
```

It is important to include the `docstring` as this allows us to make sure our code is clear. You can access that docstring using `help`:

Jupyter input

```
1  help(x_cubed)
```

```
Help on function x_cubed in module __main__:

x_cubed(x)
    A function to return x ^ 3
```

7.2.3 Call a function

Once a function is defined call it using the `()`:

Usage

```
1   name(variable1, variable2, ...)
```

For example:

Jupyter input

```
1   x_cubed(2)
```

8

Jupyter input

```
1   x_cubed(5)
```

125

Jupyter input

```
1   import sympy as sym
2
3   x = sym.Symbol("x")
4   x_cubed(x)
```

$$x^3$$

7.2.4 Run code based on a condition

To run code depending on whether or not a particular condition is met use an `if` statement.

```
if condition:
    INDENTED BLOCK OF CODE TO RUN IF CONDITION IS TRUE
else:
    OTHER INDENTED BLOCK OF CODE TO RUN IF CONDITION IS NOT TRUE
```

These `if` statements are most useful when combined with functions. For example, you can define the following function:

$$f(x) = \begin{cases} x^3 & \text{if } x < 0 \\ x^2 & \text{otherwise} \end{cases}$$

```
Jupyter input
1  def f(x):
2      """
3      A function that returns x ^ 3 if x is negative.
4      Otherwise it returns x ^ 2.
5      """
6      if x < 0:
7          return x ** 3
8      return x ** 2
```

```
Jupyter input
1  f(0)
```

0

```
Jupyter input
1  f(-1)
```

-1

```
Jupyter input
1  f(3)
```

9

Here is another example of a function that returns the price of a given item, if the item is not specific in the function then the price is 0:

```
Jupyter input
1  def get_price_of_item(item):
2      """
3      Returns the price of an item:
4
5      - 'Bread': 2
6      - 'Biscuits': 3
7      - 'Coffee': 1.80
8      - 'Tea': .50
```

```
9        - 'Toothpaste': 3.50
10
11       Other items will give a price of 0.
12       """
13       if item == "Bread":
14           return 2
15       if item == "Biscuits":
16           return 3
17       if item == "Coffee":
18           return 1.80
19       if item == "Tea":
20           return 0.50
21       if item == "Toothpaste":
22           return 3.50
23       return 0
```

Jupyter input

```
1    get_price_of_item("Toothpaste")
```

3.5

Jupyter input

```
1    get_price_of_item("Biscuits")
```

3

Jupyter input

```
1    get_price_of_item("Rollerblades")
```

0

7.2.5   Create a list using a list comprehension

You can create a new list from an old list using a **list comprehension**.

Usage

```
1    collection = [f(item) for item in iterable]
```

This corresponds to building a set from another set in the usual mathematical notation:

$$S_2 = \{f(x) \text{ for x in } S_1\}$$

If $f(x) = x - 5$ and $S_1 = \{2, 5, 10\}$, then you would have:

$$S_2 = \{-3, 0, 5\}$$

In Python this is done as follows:

Jupyter input

```
1    new_list = [object for object in old_list]
```

Jupyter input

```
1    s_1 = [2, 5, 10]
2    s_2 = [x - 5 for x in s_1]
3    s_2
```

```
[-3, 0, 5]
```

You can combine this with functions to write succinct efficient code.
For example, you can compute the price of a basket of goods using the following:

Jupyter input

```
1    basket = ["Tea", "Tea", "Toothpaste", "Bread"]
2    prices = [get_price_of_item(item) for item in basket]
3    prices
```

```
[0.5, 0.5, 3.5, 2]
```

7.2.6 Summing items in a list

You can compute the sum of items in a list using the sum tool:

Jupyter input

```
1    sum([1, 2, 3])
```

6

You can also directly use sum without specifically creating the list. This corresponds to the following mathematical notation:

$$\sum_{s \in S} f(s)$$

and is done using the following:

> ### Jupyter input
>
> ```
> 1  sum(f(object) for object in old_list)
> ```

This gives the same result as:

> ### Jupyter input
>
> ```
> 1  sum([f(object) for object in old_list])
> ```

but it is more efficient. Here is an example of getting the total price of a basket of goods:

> ### Jupyter input
>
> ```
> 1  basket = ["Tea", "Tea", "Toothpaste", "Bread"]
> 2  total_price = sum(get_price_of_item(item) for item in basket)
> 3  total_price
> ```

6.5

7.2.7 Sample from an iterable

To randomly sample from any collection of items use the random library and the `choice` tool.

> ### Usage
>
> ```
> 1  random.choice(collection)
> ```

> ### Jupyter input
>
> ```
> 1  import random
> 2
> 3  basket = ["Tea", "Tea", "Toothpaste", "Bread"]
> 4  random.choice(basket)
> ```

```
'Toothpaste'
```

7.2.8   Sample a random number

To sample a random number between 0 and 1 use the random library and the `random` tool.

Usage

```
1  random.random()
```

For example:

Jupyter input

```
1  import random
2
3  random.random()
```

0.7558634290782174

7.2.9   Reproduce random events

The random numbers processes generated by the Python random module are what are called pseudo random which means that it is possible to get a computer to reproduce them by **seeding** the random process.

Usage

```
1  random.seed(int)
```

Jupyter input

```
1  import random
2
3  random.seed(0)
4  random.random()
```

0.8444218515250481

Jupyter input

```
1  random.random()
```

0.7579544029403025

Jupyter input

```
1   random.seed(0)
2   random.random()
```

0.8444218515250481

7.3 EXERCISES

1. For each of the following, write a function, and repeatedly use it to simulate the probability of an event occurring with the following chances:

   (a) $\frac{2}{7}$

   (b) $\frac{1}{10}$

   (c) $\frac{1}{100}$

   (d) 1

2. Write a function, and repeatedly use it to simulate the probability of selecting a red token from each of the following configurations:

   (a) A bag with 4 red tokens and 4 green tokens.

   (b) A bag with 4 red tokens, 4 green tokens and 10 yellow tokens.

   (c) A bag with 0 red tokens, 4 green tokens and 10 yellow tokens.

3. An experiment consists of selecting a token from a bag and spinning a coin. The bag contains 3 red tokens and 4 blue tokens. A token is selected at random from the bag, its colour is noted and then the token is returned to the bag.

   When a red token is selected, a biased coin with probability $\frac{4}{5}$ of landing heads is spun.

   When a blue token is selected, a biased coin with probability $\frac{2}{5}$ of landing heads is spun.

   (a) Approximate the probability of picking a red token?

   (b) Approximate the probability of obtaining Heads?

   (c) If a heads is obtained, approximate the probability of having selected a red token.

4. On a randomly chose day, the probability of an individual travelling to school by car, bicycle or on foot is 1/2, 1/6 and 1/3, respectively. The probability of being late when using these methods of travel is 1/5, 2/5 and 1/10, respectively.

   (a) Approximate the probability that an individual travels by foot and is late.

   (b) Approximate the probability that an individual is not late.

   (c) Given that an individual is late, approximate the probability that they did not travel on foot.

7.4   FURTHER INFORMATION

7.4.1   What is the difference between a Python list and a Python tuple?

Two of the most used Python iterables are lists and tuples. In practice they have a number of similarities, they are both ordered collections of objects that can be used in list comprehensions as well as in other ways.

- Tuples are **immutable**

- Lists are **mutable**

This means that once created tuples cannot be changed and lists can.

As a general rule of thumb: if you do not need to modify your iterable then use a tuple as they are more computationally efficient.

7.4.2   Why does the sum of booleans count the Trues?

In the tutorial and elsewhere you created a list of booleans and then took the sum. Here are some of the steps:

```
Jupyter input
1   samples = ("Red", "Red", "Blue")
```

```
Jupyter input
1   booleans = [sample == "Red" for sample in samples]
2   booleans
```

[True, True, False]

When you take the sum of that list you get a numeric value:

```
Jupyter input
1   sum(booleans)
```

2

This has in fact counted the True values as 1 and the False values as 0.

```
Jupyter input
1   int(True)
```

1

Jupyter input

```
1   int(False)
```

0

7.4.3 What is the difference between `print` and `return`?

In functions you use the `return` statement. This does two things:

1. Assigns a value to the function run;

2. Ends the function.

The `print` statement **only** displays the output.
As an example create the following set:

$$S = \{f(x) \text{ for } x \in \{0, \pi/4, \pi/2, 3\pi/4\}\}$$

where $f(x) = \cos^2(x)$.
The correct way to do this is:

Jupyter input

```
1    import sympy as sym
2
3
4    def f(x):
5        """
6        Return the square of the cosine of x
7        """
8        return sym.cos(x) ** 2
9
10
11   S = [f(x) for x in (0, sym.pi / 4, sym.pi / 2, 3 * sym.pi / 4)]
12   S
```

```
[1, 1/2, 0, 1/2]
```

If you replaced the `return` statement in the function definition with a `print` you obtain:

```
Jupyter input
1   def f(x):
2       """
3       Return the square of the cosine of x
4       """
5       print(sym.cos(x) ** 2)
6
7
8   S = [f(x) for x in (0, sym.pi / 4, sym.pi / 2, 3 * sym.pi / 4)]
```

```
1
1/2
0
1/2
```

The function has been run and it displays the output.

However, if you look at what S is, you see that the function has not returned anything:

```
Jupyter input
1   S
```

```
[None, None, None, None]
```

7.4.4   How does Python sample randomness?

When using the Python random module, you are in fact generating a pseudo random process. True randomness is actually not common.

Pseudo randomness is an important area of mathematics as strong algorithms that create unpredictable sequences of numbers are vital to cryptographic security.

The specific algorithm used in Python for randomness is called the Mersenne twister algorithm and is state of the art.

7.4.5   What is the difference between a docstring and a comment

In Python it is possible to write statements that are ignored using the # symbol. This creates something called a "comment". For example:

```
Jupyter input
1   # create a list to represent the tokens in a bag
2   bag = ["Red", "Red", "Blue"]
```

A docstring however is something that is "attached" to a function and can be accessed by Python. If you rewrite the function to sample the experiment of the tutorial without a docstring but using comments you will have:

Jupyter input

```
1   def sample_experiment(bag):
2       # Select a token
3       selected_token = pick_a_token(container=bag)
4
5       # If the token is red then the probability of selecting heads is 2/3
6       if selected_token == "Red":
7           probability_of_selecting_heads = 2 / 3
8       # Otherwise it is 1 / 2
9       else:
10          probability_of_selecting_heads = 1 / 2
11
12      # Select a coin according to the probability.
13      if random.random() < probability_of_selecting_heads:
14          coin = "Heads"
15      else:
16          coin = "Tails"
17
18      # Return both the selected token and the coin.
19      return selected_token, coin
```

Now if you try to access the help for the function you will not get it:

Jupyter input

```
1   help(sample_experiment)
```

```
Help on function sample_experiment in module __main__:

sample_experiment(bag)
```

Furthermore, if you look at the code with comments you will see that because of the choice of variable names the comments are in fact redundant.

In software engineering it is generally accepted that comments indicate that your code is not clear and so it is preferable to write clear documentation explaining why something is done through docstrings.

Jupyter input

```
1   def sample_experiment(bag):
2       """
3       This samples a token from a given bag and then
4       selects a coin with a given probability.
5
6       If the sampled token is red then the probability
```

```
7      of selecting heads is 2/3 otherwise it is 1/2.
8
9      This function returns both the selected token
10     and the coin face.
11     """
12     selected_token = pick_a_token(container=bag)
13
14     if selected_token == "Red":
15         probability_of_selecting_heads = 2 / 3
16     else:
17         probability_of_selecting_heads = 1 / 2
18
19     if random.random() < probability_of_selecting_heads:
20         coin = "Heads"
21     else:
22         coin = "Tails"
23     return selected_token, coin
```

CHAPTER 8

Sequences

The formal definition of sequences is a collection of ordered objects with potential repetitions. The study of these sequences leads to many interesting results. Here you will concentrate on using recursive definitions to generate the values in a sequence.

In this chapter you will cover:

- Using recursion .

8.1 TUTORIAL

You will solve the following problem using a computer a programming technique called **recursion**.

A sequence $a_1, a_2, a_3, \ldots$ is defined by:

$$\begin{cases} a_1 = k, \\ a_{n+1} = 2a_n - 7, n \geq 1, \end{cases}$$

where k is a constant.

1. Write down an expression for a_2 in terms of k.

2. Show that $a_3 = 4k - 21$

3. Given that $\sum_{r=1}^{4} a_r = 43$ find the value of k.

You will use Python to define a function that reproduces the mathematical definition of a_k:

Jupyter input

```
def generate_a(k_value, n):
    """
    Uses recursion to return a_n for a given value of k:

    a_1 = k
    a_n = 2a_{n-1} - 7
    """
    if n == 1:
```

 DOI: 10.1201/9781003451860-8

```
9        return k_value
10    return 2 * generate_a(k_value, n - 1) - 7
```

This is similar to the mathematical definition: the Python definition of the function refers to itself.

You can use this to compute a_3 for $k = 4$:

Jupyter input

```
1  generate_a(k_value=4, n=3)
```

-5

You can use this to compute a_5 for $k = 1$:

Jupyter input

```
1  generate_a(k_value=1, n=5)
```

-89

Finally it is also possible to pass a symbolic value to k_value. This allows you to answer the first question:

Jupyter input

```
1  import sympy as sym
2
3  k = sym.Symbol("k")
4  generate_a(k_value=k, n=2)
```

$$2k - 7$$

Likewise for a_3:

Jupyter input

```
1  generate_a(k_value=k, n=3)
```

$$4k - 21$$

For the last question start by computing the sum:

$$\sum_{r=1}^{4} a_r$$

Jupyter input

```
1  sum_of_first_four_terms = sum(generate_a(k_value=k, n=r) for r in
   ↪  range(1, 5))
2  sum_of_first_four_terms
```

$$15k - 77$$

This allows you to create the given equation and solve it:

Jupyter input

```
1  equation = sym.Eq(sum_of_first_four_terms, 43)
2  sym.solveset(equation, k)
```

$$\{8\}$$

In this tutorial you have

- Defined a function using recursion.

- Called this function using both numeric and symbolic values.

8.2   HOW TO

8.2.1   Define a function using recursion

It is possible to define a recursive expression using a recursive function in Python. This requires two things:

- A recursive rule: something that relates the current term to another one;

- A base case: a particular term that does not need the recursive rule to be calculated.

Consider the following mathematical expression:

$$\begin{cases} a_1 = 1, \\ a_n = 2a_{n-1}, n > 1, \end{cases}$$

- The recursive rule is: $a_n = 2a_{n-1}$;

- The base case is: $a_1 = 1$.

In Python this can be written as:

Jupyter input

```python
def generate_sequence(n):
    """
    Generate the sequence defined by:

    a_1 = 1
    a_n = 2 a_{n - 1}

    This is done using recursion.
    """
    if n == 1:
        return 1
    return 2 * generate_sequence(n - 1)
```

Here you can get the first seven terms:

Jupyter input

```python
values_of_sequence = [generate_sequence(n) for n in range(1, 8)]
values_of_sequence
```

```
[1, 2, 4, 8, 16, 32, 64]
```

8.3 EXERCISES

1. Using recursion, obtain the first ten terms of the following sequences:

   (a) $\begin{cases} a_1 = 1, \\ a_n = 3a_{n-1}, n > 1 \end{cases}$

   (b) $\begin{cases} b_1 = 3, \\ b_n = 6b_{n-1}, n > 1 \end{cases}$

   (c) $\begin{cases} c_1 = 3, \\ c_n = 6c_{n-1} + 3, n > 1 \end{cases}$

   (d) $\begin{cases} d_0 = 3, \\ d_n = \sqrt{d_{n-1}} + 3, n > 0 \end{cases}$

2. Using recursion, obtain the first five terms of the sequence:

$$\begin{cases} a_0 = 0, \\ a_1 = 1, \\ a_n = a_{n-1} + a_{n-2}, n \geq 2 \end{cases}$$

3. A 40-year building programme for new houses began in Oldtown in the year 1951 (Year 1) and finished in 1990 (Year 40).

   The number of houses built each year form an arithmetic sequence with first term a and common difference d.

   Given that 2400 new houses were built in 1960 and 600 new houses were built in 1990, find:

   (a) The value of d.

   (b) The value of a.

   (c) The total number of houses built in Oldtown over 40 years.

4. A sequence is given by:

$$\begin{cases} x_1 = 1 \\ x_{n+1} = x_n(p + x_n), n > 1 \end{cases}$$

   for $p \neq 0$.

   (a) Find x_2 in terms of p.

   (b) Show that $x_3 = 1 + 3p + 2p^2$.

   (c) Given that $x_3 = 1$, find the value of p

8.4   FURTHER INFORMATION

8.4.1   What are the differences between recursion and iteration?

When giving instructions to a computer it is possible to use recursion to directly implement a common mathematical definition. For example consider the following sequence:

$$\begin{cases} a_1 = 1 \\ a_{n+1} = 3a_n, n > 1 \end{cases}$$

You can define this in Python as follows:

Jupyter input

```
def generate_sequence(n):
    """
    Generate the sequence defined by:

    a_1 = 1
    a_n = 3 a_{n - 1}

    This is done using recursion.
    """
    if n == 1:
        return 1
    return 3 * generate_sequence(n - 1)
```

The first six terms:

Jupyter input

```
1   [generate_sequence(n) for n in range(1, 7)]
```

`[1, 3, 9, 27, 81, 243]`

In this case this corresponds to powers of 3, and indeed you can prove that: $a_n = 3^{n-1}$. The proof is not given here but one approach to doing it would be to use induction which is closely related to recursive functions.

You can write a different python function that uses this formula. This is called **iteration**:

Jupyter input

```
1   def calculate_sequence(n):
2       """
3       Calculate the nth term of the sequence defined by:
4
5       a_1 = 1
6       a_n = 3 a_{n - 1}
7
8       This is done using iteration using the direct formula:
9
10      a_n = 3 ^ n
11      """
12      return 3 ** (n - 1)
```

Jupyter input

```
1   [calculate_sequence(n) for n in range(1, 7)]
```

`[1, 3, 9, 27, 81, 243]`

You can in fact use a Jupyter command to time the run time of a command. It is clear that recursion is slower.

Jupyter input

```
1   %timeit [generate_sequence(n) for n in range(1, 25)]
```

`19.2 µs ± 246 ns per loop (mean ± std. dev. of 7 runs, 100,000 loops each)`

Jupyter input

```
1  %timeit [calculate_sequence(n) for n in range(1, 25)]
```

5.63 µs ± 44.7 ns per loop (mean ± std. dev. of 7 runs, 100,000 loops each)

In practice:

- Using recursion is powerful as it can be used to directly implement recursive definitions.

- Using iteration is more computationally efficient but it is not always straightforward to obtain an iterative formula.

8.4.2   What is caching?

One of the reasons that recursion is computationally inefficient is that it always has to recalculate previously calculated values.

For example:

$$\begin{cases} a_1 = 1 \\ a_{n+1} = 3a_n, n > 1 \end{cases}$$

One way to address this is by using caching, where a function stores the result of a computation so that if it is called again with the same input, it can quickly retrieve the stored value instead of recalculating it. Python has a caching tool available in the functools library:

Jupyter input

```
1   import functools
2
3
4   def generate_sequence(n):
5       """
6       Generate the sequence defined by:
7
8       a_1 = 1
9       a_n = 3 a_{n - 1}
10
11      This is done using recursion.
12      """
13      if n == 1:
14          return 1
15      return 3 * generate_sequence(n - 1)
16
17
18  @functools.lru_cache()
19  def cached_generate_sequence(n):
```

```
20          """
21          Generate the sequence defined by:
22
23          a_1 = 1
24          a_n = 3 a_{n - 1}
25
26          This is done using recursion but also includes a cache.
27          """
28          if n == 1:
29              return 1
30          return 3 * cached_generate_sequence(n - 1)
```

Timing both these approaches confirms a substantial increase in time for the cached version.

Jupyter input

```
1    %timeit [generate_sequence(n) for n in range(1, 25)]
```

20.5 µs ± 381 ns per loop (mean ± std. dev. of 7 runs, 100,000 loops each)

Jupyter input

```
1    %timeit [cached_generate_sequence(n) for n in range(1, 25)]
```

---

934 ns ± 38.1 ns per loop (mean ± std. dev. of 7 runs, 1,000,000 loops each)

CHAPTER 9

Statistics

Statistics is described as:

"Statistics is the discipline that concerns the collection, organization, analysis, interpretation, and presentation of data."

In practice this often means doing some form of analysis of data. This includes processes like taking a mean of a collection of numerical values and checking if particular relationship exists within the data.

In this chapter you will cover:

- Calculating measures of central tendency and spread;

- Calculating bivariate coefficients;

- Fitting a line of best fit; and

- Using the Normal distribution.

9.1 TUTORIAL

You will solve the following problem using a computer to do some of the more tedious calculations.

Anna is investigating the relationship between exercise and resting heart rate. She takes a random sample of 19 people in her year group and records for each person

- their resting heart rate, h beats per minute.

- the number of minutes, m, spent exercising each week.

Table 9.1 shows the data.
You can see a scatter plot in Figure 9.1.

1. For all collected values of h and m obtain:

    - The mean
    - The median
    - The quartiles
    - The standard deviation
    - The variation

DOI: 10.1201/9781003451860-9

TABLE 9.1 Data Showing Collected
Samples of Heart Rate (h) and Weekly
Minutes of Exercise (m)

h	m
76.0	5
72.0	5
71.0	21
74.0	30
71.0	42
69.0	20
68.0	20
68.0	35
66	80.0
64	120.0
65	140.0
63	180.0
63	205.0
62	225.0
65	237.0
63	280.0
65	300.0
64	356.0
64	360.0

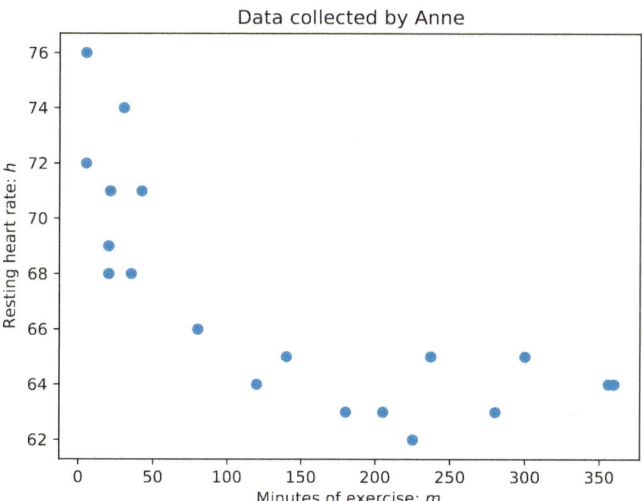

Figure 9.1 A scatter plot of the data collected by Anna.

- The maximum

- The minimum

2. Obtain the Pearson Coefficient of correlation for the variables h and m.

3. Obtain the line of best fit for variables x and y as defined by:

$$x = \ln(m) \qquad y = \ln(h)$$

4. Using the above obtain a relationship between m and h of the form:

$$h = cm^k$$

Start by inputting all the data:

```
 1  h = (
 2      76.0,
 3      72.0,
 4      71.0,
 5      74.0,
 6      71.0,
 7      69.0,
 8      68.0,
 9      68.0,
10      66.0,
11      64.0,
12      65.0,
13      63.0,
14      63.0,
15      62.0,
16      65.0,
17      63.0,
18      65.0,
19      64.0,
20      64.0,
21  )
22  m = (
23      5,
24      5,
25      21,
26      30,
27      42,
28      20,
29      20,
30      35,
31      80,
32      120,
```

```
33        140,
34        180,
35        205,
36        225,
37        237,
38        280,
39        300,
40        356,
41        360,
42    )
```

The main tool you are going to use for this problem is `statistics`.

Jupyter input

```
1    import statistics as st
```

To calculate the mean:

Jupyter input

```
1    st.mean(h)
```

67.0

Jupyter input

```
1    st.mean(m)
```

140.05263157894737

To calculate the median:

Jupyter input

```
1    st.median(h)
```

65.0

Jupyter input

```
1  st.median(h)
```

```
120
```

To calculate the quartiles, use `statistics.quantiles` and specify that you want to separate the data into $n = 4$ quarters.

Jupyter input

```
1  st.quantiles(h, n=4)
```

```
[64.0, 65.0, 71.0]
```

Jupyter input

```
1  st.quantiles(m, n=4)
```

```
[21.0, 120.0, 237.0]
```

Note that this calculation confirms the median which corresponds to the 50% quartile. To calculate the sample standard deviation:

Jupyter input

```
1  st.stdev(h)
```

```
4.123105625617661
```

Jupyter input

```
1  st.stdev(m)
```

```
124.46662813970593
```

To calculate the sample variance:

Jupyter input

```
1  st.variance(h)
```

```
17.0
```

Jupyter input

```
1  st.variance(m)
```

15491.941520467837

To compute the maximum:

Jupyter input

```
1  max(h)
```

76.0

Jupyter input

```
1  max(m)
```

360

To compute the minimum:

Jupyter input

```
1  min(h)
```

62.0

Jupyter input

```
1  min(m)
```

5

To compute the Pearson Coefficient of correlation use `statistics.correlation`:

Jupyter input

```
1  st.correlation(h, m)
```

-0.7686142969026402

This negative value indicates a negative correlation between h and m, indicating that the more you exercise the lower your heart rate is likely to be. To calculate the line of best fit for the transformed variables you need to first create them. You will do this using a list comprehension. As you are doing everything numerically, you will use `math.log` which by default computes the natural logarithm.

Jupyter input

```python
import math
x = [math.log(value) for value in m]
y = [math.log(value) for value in h]
```

Now to compute the line of best fit use `statistics.linear_regression`:

Jupyter input

```python
slope, intercept = st.linear_regression(x, y)
```

The slope is:

Jupyter input

```python
slope
```

```
-0.03854770754231997
```

The intercept is:

Jupyter input

```python
intercept
```

```
4.368415819445762
```

Recall the transformation of the variables:

$$x = \ln(m) \qquad y = \ln(h)$$

You now have the relationship:

$$y = ax + b$$

Where a corresponds to the `slope` and b corresponds to the `intercept`. The question asks for a relationship between m and h of the form:

$$h = cm^k$$

You can use `sympy` to manipulate the expressions:

```
1  import sympy as sym
2
3  h = sym.Symbol("h")
4  m = sym.Symbol("m")
5  a = sym.Symbol("a")
6  b = sym.Symbol("b")
7  x = sym.ln(m)
8  y = sym.ln(h)
```

A general expression for x and y can be expressed in terms of m and h:

```
1  line = sym.Eq(lhs=y, rhs=a * x + b)
2  line
```

$$\log{(h)} = a \log{(m)} + b$$

Taking the exponential of both sides gives the required relationship:

```
1  sym.exp(line.lhs)
```

$$h$$

```
1  sym.expand(sym.exp(line.rhs))
```

$$e^b e^{a \log{(m)}}$$

Which can be rewritten as:

$$e^b m^a$$

Substituting the values for the slope and intercept into these expressions gives the required relationship:

```
1   sym.exp(line.rhs).subs({a: slope, b: intercept})
```

$$\frac{78.9185114479915}{m^{0.03854770754232}}$$

Figure 9.2 is a plot that shows this relationship.

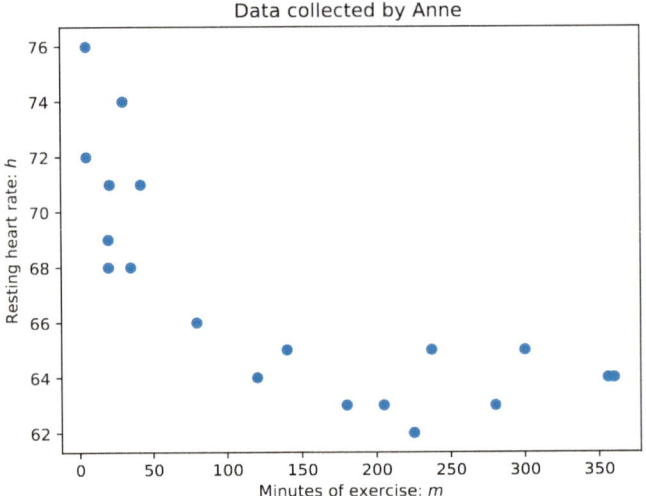

Figure 9.2 A scatter plot of the data collected by Anne with the fitted relationship.

In this tutorial you have

- Calculated values of central tendency and spread;

- Calculated some bivariate coefficients;

- Fitted a line of best fit.

9.2 HOW TO

9.2.1 Calculate measures of spread and tendency

9.2.1.1 Calculate a mean

You can calculate the mean of a set of data using `statistics.mean` which takes an iterable.

Usage

```
1   statistics.mean(data)
```

For example, to calculate the mean of $(1, 5, 10, 12, 13, 20)$:

Jupyter input

```
1   import statistics as st
2
3   data = (1, 5, 10, 12, 13, 20)
4   st.mean(data)
```

```
10.166666666666666
```

9.2.1.2 Calculate a median

You can calculate the median of a set of data using `statistics.median` which takes an iterable.

Usage

```
1   statistics.median(data)
```

For example, to calculate the median of $(1, 5, 10, 12, 13, 20)$:

Jupyter input

```
1   import statistics as st
2
3   data = (1, 5, 10, 12, 13, 20)
4   st.median(data)
```

```
11.0
```

9.2.1.3 Calculate the population standard deviation

You can calculate the population standard deviation of a set of data using `statistics.pstdev` which takes an iterable.

Usage

```
1   statistics.pstdev(data)
```

For example, to calculate the population standard deviation of $(1, 5, 10, 12, 13, 20)$:

Jupyter input

```
1   import statistics as st
2
3   data = (1, 5, 10, 12, 13, 20)
4   st.pstdev(data)
```

```
6.039223643997813
```

9.2.1.4 Calculate the sample standard deviation

You can calculate the sample standard deviation of a set of data using `statistics.stdev` which takes an iterable.

Usage

```
1   statistics.stdev(data)
```

For example, to calculate the sample standard deviation of $(1, 5, 10, 12, 13, 20)$:

Jupyter input

```
1   import statistics as st
2
3   data = (1, 5, 10, 12, 13, 20)
4   st.stdev(data)
```

```
6.6156380392723015
```

9.2.1.5 Calculate the population variance

You can calculate the population variance of a set of data using `statistics.pvariance` which takes an iterable.

Usage

```
1   statistics.pvariance(data)
```

For example, to calculate the population variance of $(1, 5, 10, 12, 13, 20)$:

Jupyter input

```
1  import statistics as st
2
3  data = (1, 5, 10, 12, 13, 20)
4  st.pvariance(data)
```

36.47222222222222

9.2.1.6 Calculate the sample variance

You can calculate the sample variance of a set of data using `statistics.variance` which takes an iterable.

Usage

```
1  statistics.variance(data)
```

For example, to calculate the sample variance of $(1, 5, 10, 12, 13, 20)$:

Jupyter input

```
1  import statistics as st
2
3  data = (1, 5, 10, 12, 13, 20)
4  st.variance(data)
```

43.766666666666666

9.2.1.7 Calculate the maximum

You can calculate the maximum of a set of data using `max` which takes an iterable:

Usage

```
1  max(data)
```

For example, to calculate the maximum of $(1, 5, 10, 12, 13, 20)$:

Jupyter input

```
1  data = (1, 5, 10, 12, 13, 20)
2  max(data)
```

```
20
```

9.2.1.8 Calculate the minimum

You can calculate the minimum of a set of data use `min` which takes an iterable:

> **Usage**
> ```
> 1   min(data)
> ```

For example, to calculate the minimum of $(1, 5, 10, 12, 13, 20)$:

> **Jupyter input**
> ```
> 1   data = (1, 5, 10, 12, 13, 20)
> 2   min(data)
> ```

```
1
```

9.2.1.9 Calculate quantiles

To calculate cut points dividing data into n intervals of equal probability you can use `statistics.quantiles` which takes an iterable and a number of intervals.

> **Usage**
> ```
> 1   statistics.quantiles(data, n)
> ```

For example, to calculate the cut points that divide $(1, 5, 10, 12, 13, 20)$ into four intervals of equal probability (in this case the quantiles are called quartiles):

> **Jupyter input**
> ```
> 1   import statistics as st
> 2
> 3   data = (1, 5, 10, 12, 13, 20)
> 4   st.quantiles(data, n=4)
> ```

```
[4.0, 11.0, 14.75]
```

9.2.2 Calculate the sample covariance

To calculate the sample covariance of two data sets you can use `statistics.covariance` which takes two iterables.

```
1  statistics.covariance(first_data_set, second_data_set)
```

For example, to calculate the sample covariance of $x = (1, 5, 10, 12, 13, 20)$ and $y = (3, -3, 6, -2, 1, 2)$:

Jupyter input

```
1  import statistics as st
2
3  x = (1, 5, 10, 12, 13, 20)
4  y = (3, -3, 6, -2, 1, 2)
5  st.covariance(x, y)
```

```
1.1666666666666674
```

9.2.3 Calculate the Pearson correlation coefficient

To calculate the correlation coefficient of two data sets you can use `statistics.correlation` which takes two iterables.

Usage

```
1  statistics.correlation(first_data_set, second_data_set)
```

For example, to calculate the correlation coefficient of $x = (1, 5, 10, 12, 13, 20)$ and $y = (3, -3, 6, -2, 1, 2)$:

Jupyter input

```
1  import statistics as st
2
3  x = (1, 5, 10, 12, 13, 20)
4  y = (3, -3, 6, -2, 1, 2)
5  st.correlation(x, y)
```

```
0.05325222181462787
```

9.2.4 Fit a line of best fit

To carry out linear regression to fit a line of best fit between two data sets you can use `statistics.linear_regression` which takes two iterables and returns a tuple with the slope and the intercept of the line.

Usage

```
1  statistics.linear_regression(first_data_set, second_data_set)
```

For example, to calculate the correlation coefficient of $x = (1, 5, 10, 12, 13, 20)$ and $y = (-3, -14, -31, -6, -40, -70)$:

Jupyter input

```
1  import statistics as st
2
3  x = (1, 5, 10, 12, 13, 20)
4  y = (-3, -14, -31, -6, -40, -70)
5  st.linear_regression(x, y)
```

```
LinearRegression(slope=-3.2338156892612333, intercept=5.543792840822537)
```

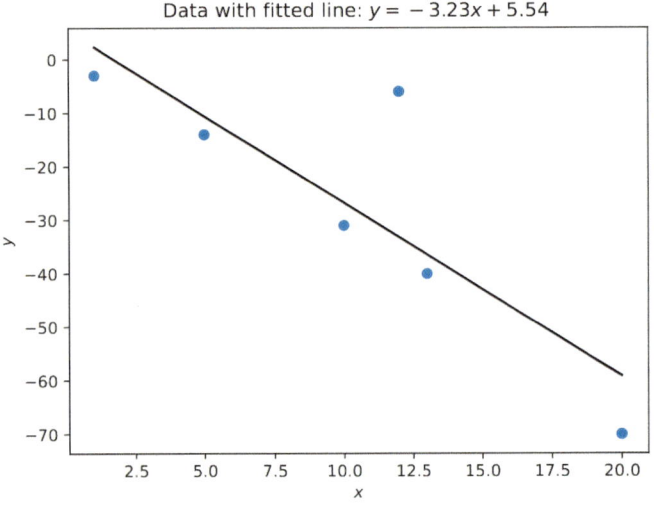

Figure 9.3   A line of best fit.

9.2.5   Create an instance of the normal distribution

A normal distribution with mean μ and standard deviation σ can be created using statistics.NormalDist:

Usage

```
1  statistics.NormalDist(mu, sigma)
```

For example, to create the normal distribution with $\mu = 3$ and $\sigma = .5$:

Jupyter input

```
1  import statistics as st
2
3  distribution = st.NormalDist(mu=3, sigma=.5)
4  distribution
```

```
NormalDist(mu=3.0, sigma=0.5)
```

9.2.6 Use the cumulative distribution function of a normal distribution

For an instance of a normal distribution with mean μ and σ, the cumulative distribution function which gives $F(x) = P(X < x)$ (the probability that the normally distributed random variable is less than X) can be accessed using `statistics.NormaDist.cdf`.

Usage

```
1  distribution = statistics.NormalDist(mu, sigma)
2  distribution.cdf(x)
```

For example, to find the probability that $X < 2$ for a normally distributed random variable with $\mu = 3$ and $\sigma = .5$:

Jupyter input

```
1  import statistics as st
2
3  distribution = st.NormalDist(mu=3, sigma=.5)
4  distribution.cdf(2)
```

```
0.02275013194817921
```

9.2.7 Use the inverse cumulative distribution function of a normal distribution

For an instance of a normal distribution with mean μ and σ, the inverse cumulative distribution function which for a given p gives x such that $p = P(X < x)$ can be accessed using `statistics.NormaDist.inv_cdf`.

Usage

```
1  distribution = statistics.NormalDist(mu, sigma)
2  distribution.inv_cdf(p)
```

For example, to find the value of X for which a normally distributed random variable with $\mu = 3$ and $\sigma = .5$ will be less than with probability .7.

Jupyter input

```
1  import statistics as st
2
3  distribution = st.NormalDist(mu=3, sigma=.5)
4  distribution.inv_cdf(.7)
```

3.2622002563540202

9.3 EXERCISES

1. For each of the following sets of data:

(a) Data set 1:

Jupyter input

```
1   data_set_1 = (
2        74,
3        -7,
4        58,
5        82,
6        60,
7        3,
8        49,
9        85,
10       24,
11       99,
12       73,
13       76,
14       11,
15       -4,
16       61,
17       87,
18       93,
19       13,
20       1,
```

```
21      28,
22  )
```

(b) Data set 2:

```
Jupyter input

1  data_set_2 = (
2      65,
3      59,
4      81,
5      81,
6      76,
7      93,
8      91,
9      88,
10     55,
11     97,
12     86,
13     94,
14     79,
15     54,
16     63,
17     56,
18     58,
19     77,
20     85,
21     88,
22  )
```

(c) Data set 3:

```
Jupyter input

1  data_set_3 = (
2      0.31,
3      -0.13,
4      0.19,
5      0.46,
6      -0.27,
7      -0.06,
8      0.20,
9      0.42,
10     -0.07,
11     0.11,
```

```
12        -0.11,
13        -0.43,
14        -0.36,
15         0.45,
16        -0.42,
17         0.11,
18         0.08,
19         0.31,
20         0.48,
21         0.17,
22     )
```

(d) Data set 4:

Jupyter input

```
1    data_set_4 = (
2         2,
3         4,
4         2,
5         2,
6         2,
7         2,
8         2,
9         3,
10        2,
11        2,
12        2,
13        4,
14        2,
15        4,
16        2,
17        2,
18        3,
19        4,
20        3,
21        4,
22    )
```

Calculate:

- The mean,
- The median,
- The max,
- The min,
- The population standard deviation,
- The sample standard deviation,

- The population variance,
- The sample variance,
- The quartiles (the set of $n = 4$ quantiles), and
- The deciles (the set of $n = 10$ quantiles).

2. Calculate the sample covariance and the correlation coefficient for the following pairs of data sets from question 1:

    (a) data_set_1 and data_set_4
    (b) data_set_3 and data_set_4
    (c) data_set_2 and data_set_3
    (d) data_set_1 and data_set_2

3. For each of the data sets from question 1 obtain the covariance and correlation coefficient for the data set with itself.

4. Obtain a for the pairs of data sets from question 2.

5. Given a collection of 250 individuals whose height is normally distributed with mean 165 and standard deviation 5. What is the expected number of individuals with height between 150 and 160?

6. Consider a class test where the scores are normally distributed with mean 65 and standard deviation 5.

    (a) What is the probability of failing the class test (a score less than 40)?
    (b) What proportion of the class gets a first class mark (a score above 70)?
    (c) What is the mark that only 10% of the class would expect to get more than?

9.4 FURTHER INFORMATION

9.4.1 What is the difference between the sample and the population variance and standard deviation?

For a given set of N values $x_1, x_2, \ldots, x_N$ with mean $\bar{x}$ the sample standard deviation is given by:

$$\sigma = \sqrt{\frac{\sum_{i=1}^{N} (x_i - \bar{x})^2}{N - 1}}$$

The sample variance is given by:

$$\sigma^2$$

The population standard deviation is given by:

$$\sigma = \sqrt{\frac{\sum_{i=1}^{N} (x_i - \bar{x})^2}{N}}$$

The population variance is given by:

$$\sigma^2$$

The population standard deviation and/or variance should be used when the data set in question is for the entire population.

The sample standard deviation and/or variance should be used when the data set in question is a sample of the entire population. The modification in the calculation is to counteract a potential bias.

9.4.2 How to plot a line of best fit?

The main library for plotting is called `matplotlib`. Below is some code to plot the data and regression line for two collections of data. Figure 9.4 gives the output.

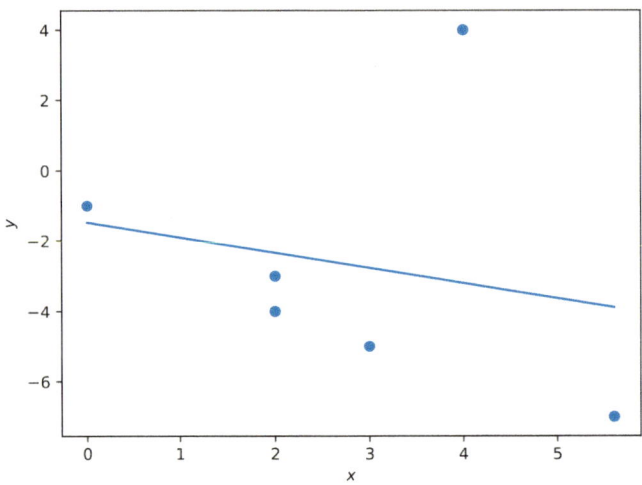

Figure 9.4 Example of plotting a fitted line.

```
1   import statistics as stat
2   import matplotlib.pyplot as plt
3
4   x = (0, 2, 2, 3, 4, 5.6)
5   y = (-1, -3, -4, -5, 4, -7)
6
7   slope, intercept = stat.linear_regression(x, y)
8
9   start_point, end_point = min(x), max(x)
10  image_start_point = slope * start_point + intercept
11  image_end_point = slope * end_point + intercept
12
13  plt.figure()
14  plt.scatter(x, y)
```

Jupyter input

```
15  plt.plot((start_point, end_point), (image_start_point,
    ↪  image_end_point))
16  plt.xlabel("$x$")
17  plt.ylabel("$y$")
```

9.4.3 What other statistics tools exist in Python?

The statsmodels library allows for a wider breadth of statistical analysis. The scikit-learn library is arguably one of the most popular python libraries. It is technically a library for machine learning and not statistics.

9.4.4 What is the difference between machine learning and statistics?

In a lot of cases the difference here is more question of vocabulary than actual tangible differences.

For example, the scikit-learn library has a tool for linear regression as does the statsmodels and the statistics library.

In practice statistics is often more descriptive, for example, using linear regression to understand the relationship between two variables. Whereas machine learning is more predictive, for example using liner regression to predict one variable value from another.

A lot of modern applied mathematics uses tools such as neural networks which are considered to be from the field of machine learning.

CHAPTER 10

Differential Equations

A differential equation is an equation that relates one or more quantities and their derivatives. These can often be used to model real-world systems.

In this chapter you will cover:

- Creating a symbolic function;

- Writing a differential equation; and

- Solving a differential equation.

10.1 TUTORIAL

You will solve the following problem using a computer to do some of the more tedious calculations.

A container has volume V of liquid which is poured in at a rate proportional to e^{-t} (where t is some measurement of time). Initially the container is empty and after $t = 3$ time units the volume of liquid is 15.

1. Show that $V(t) = \frac{-15e^3}{1-e^3}(1 - e^{-t})$.

2. Obtain the limit $\lim_{t\to\infty} V(t)$.

You first need to create the differential equation described in the question:

Jupyter input

```
1  import sympy as sym
2
3  t = sym.Symbol("t")
4  k = sym.Symbol("k")
5  V = sym.Function("V")
6
7  differential_equation = sym.Eq(lhs=sym.diff(V(t), t), rhs=k *
   ↪  sym.exp(-t))
8  differential_equation
```

DOI: 10.1201/9781003451860-10

$$\frac{d}{dt}V(t) = ke^{-t}$$

In order to solve the differential equation write:

Jupyter input

```
1  sym.dsolve(differential_equation, V(t))
```

$$V(t) = C_1 - ke^{-t}$$

Note that the question gives an initial condition: "initially the container is empty" which corresponds to $V(0) = 0$. You can pass this to the call to solve the differential equation:

Jupyter input

```
1  condition = {V(0): 0}
2  particular_solution = sym.dsolve(differential_equation, V(t),
   ↪ ics=condition)
3  sym.simplify(particular_solution)
```

$$V(t) = k - ke^{-t}$$

You also know that $V(3) = 15$ which corresponds to the following equation:

Jupyter input

```
1  equation = sym.Eq(particular_solution.rhs.subs({t: 3}), 15)
2  equation
```

$$-\frac{k}{e^3} + k = 15$$

You can solve this equation to find a value for k:

Jupyter input

```
1  sym.simplify(sym.solveset(equation, k))
```

$$\left\{ -\frac{15e^3}{1 - e^3} \right\}$$

which is the required value.

You can use the complete expression for $V(t)$ to take the limit:

> **Jupyter input**
>
> ```
> 1  limit = sym.limit((-15 * sym.exp(3) / (1- sym.exp(3))) * (1 -
>    ↪  sym.exp(-t)), t, sym.oo)
> 2  limit
> ```

$$-\frac{15e^3}{1 - e^3}$$

This is approximately:

> **Jupyter input**
>
> ```
> 1  float(limit)
> ```

```
15.78593544736884
```

> In this tutorial you have
>
> - Created a differential equation.
>
> - Obtained the general solution of a differential equation.
>
> - Obtained the particular solution of a differential equation.

10.2   HOW TO

10.2.1   Create a symbolic function

To create a symbolic function use `sympy.Function`.

> **Usage**
>
> ```
> 1  sympy.Function("y")
> ```

For example:

> **Jupyter input**
>
> ```
> 1  import sympy as sym
> 2
> 3  y = sym.Function("y")
> 4  y
> ```

y

You can pass symbolic variables to this symbolic function:

> **Jupyter input**
>
> ```
> 1  x = sym.Symbol("x")
> 2  y(x)
> ```

$$y(x)$$

Now, you can create the derivative of a symbolic function:

> **Jupyter input**
>
> ```
> 1  sym.diff(y(x), x)
> ```

$$\frac{d}{dx}y(x)$$

10.2.2 Create a differential equation

To create a differential equation use `sympy.Eq`.

> **Usage**
>
> ```
> 1  import sympy as sym
> 2
> 3  y = sym.Function("y")
> 4  x = sym.Symbol("x")
> 5
> 6  equation = sym.Eq(lhs, rhs)
> ```

Where `lhs` and `rhs` are expressions in y, $\frac{dy}{dx}$ and x.
For example, to create the differential equation: $\frac{dy}{dx} = \cos(x)y$ write:

Jupyter input

```
1  import sympy as sym
2
3  y = sym.Function("y")
4  x = sym.Symbol("x")
5
6  lhs = sym.diff(y(x), x)
7  rhs = sym.cos(x) * y(x)
8  differential_equation = sym.Eq(lhs, rhs)
9  differential_equation
```

$$\frac{d}{dx} y(x) = y(x) \cos(x)$$

10.2.3 Obtain the general solution of a differential equation

To obtain the generation solution to a differential equation use: `sympy.dsolve`.

Usage

```
1  import sympy as sym
2
3  y = sym.Function("y")
4  x = sym.Symbol("x")
5
6  equation = sym.Eq(lhs, rhs)
7  sym.dsolve(equation, y(x))
```

For example, to solve the differential equation: $\frac{dy}{dx} = \cos(x)y$ write:

Jupyter input

```
1  import sympy as sym
2
3  y = sym.Function("y")
4  x = sym.Symbol("x")
5
6  lhs = sym.diff(y(x), x)
7  rhs = sym.cos(x) * y(x)
8  differential_equation = sym.Eq(lhs, rhs)
9  sym.dsolve(differential_equation, y(x))
```

$$y(x) = C_1 e^{\sin(x)}$$

10.2.4 Obtain the particular solution of a differential equation

To obtain the particular solution to a differential equation use: `sympy.dsolve` and pass the initial conditions: `ics`.

Usage

```
1  import sympy as sym
2
3  y = sym.Function("y")
4  x = sym.Symbol("x")
5
6  equation = sym.Eq(lhs, rhs)
7  sym.dsolve(equation, y(x), ics={y(x_0): value})
```

For example, to solve the differential equation: $\frac{dy}{dx} = \cos(x)y$ with the condition $y(5) = \pi$ write:

Jupyter input

```
1  import sympy as sym
2
3  y = sym.Function("y")
4  x = sym.Symbol("x")
5
6  lhs = sym.diff(y(x), x)
7  rhs = sym.cos(x) * y(x)
8  differential_equation = sym.Eq(lhs, rhs)
9
10  condition = {y(5): sym.pi}
11  sym.dsolve(differential_equation, y(x), ics=condition)
```

$$y(x) = \pi e^{-\sin(5)} e^{\sin(x)}$$

The syntax used here is similar to substituting values into algebraic expressions (see Section 3.3.7).

10.3 EXERCISES

1. Create the following differential equations:

   (a) $\frac{dy}{dx} = \cos(x)$

   (b) $\frac{dy}{dx} = 1 - y$

   (c) $\frac{dy}{dx} = \frac{x - 50}{10}$

(d) $\frac{dy}{dx} = y^2 \ln(x)$

(e) $\frac{dy}{dx} = (1 + y)^2$

2. Obtain the general solution for the equations in question 1.

3. Obtain the particular solution for the equations in question 1 with the following particular conditions:

   (a) $y(0) = \pi$

   (b) $y(2) = 3$

   (c) $y(50) = 1$

   (d) $y(e) = 1$

   (e) $y(-1) = 3$

4. The rate of increase of a population (p) is equal to 1% of the size of the population.

   (a) Define the differential equation that models this situation.

   (b) Given that $p(0) = 5000$ find the population after 5 time units.

5. The rate of change of the temperature of a hot drink is proportional to the difference between the temperature of the drink (T) and the room temperature (T_R).

   (a) Define the differential equation that models this situation.

   (b) Solve the differentia equation.

   (c) Given that $T(0) = 100$ and the room temperature is $T_R = 20$ obtain the particular solution.

   (d) Use the particular solution to identify how on it will take for the drink to be ready for consumption (a temperature of 80) given that after 3 time units $T(3) = 90$.

10.4 FURTHER INFORMATION

10.4.1 How to solve a system of differential equations?

Given a system of differential equations like the following:

$$\begin{cases} \frac{dx}{dt} = & x - y \\ \frac{dy}{dt} = & x + y \\ y(0) = & 250 \\ y(1) = & 300 \end{cases}$$

You can solve it using `sym.dsolve` but instead of passing a single differential equation, pass an iterable of multiple equations:

Jupyter input

```
1   import sympy as sym
2
3
4   y = sym.Function("y")
5   x = sym.Function("x")
6
7   t = sym.Symbol("t")
8   alpha = sym.Symbol("alpha")
9   beta = sym.Symbol("beta")
10
11  system_of_equations = (
12      sym.Eq(sym.diff(y(t), t), alpha * x(t)),
13      sym.Eq(sym.diff(x(t), t), beta * y(t)),
14  )
15  conditions = {y(0): 250, y(1): 300}
16
17  y_solution, x_solution = sym.dsolve(system_of_equations,
    ↪  ics=conditions, set=True)
18  x_solution
```

$$x(t) = -\frac{50\beta\left(5e^{\sqrt{\alpha\beta}} - 6\right)e^{\sqrt{\alpha\beta}}e^{-t\sqrt{\alpha\beta}}}{\sqrt{\alpha\beta}\left(e^{2\sqrt{\alpha\beta}} - 1\right)} + \frac{50\beta\left(6e^{\sqrt{\alpha\beta}} - 5\right)e^{t\sqrt{\alpha\beta}}}{\sqrt{\alpha\beta}\left(e^{2\sqrt{\alpha\beta}} - 1\right)}$$

Jupyter input

```
1   y_solution
```

$$y(t) = \frac{50 \cdot \left(5e^{\sqrt{\alpha\beta}} - 6\right)e^{\sqrt{\alpha\beta}}e^{-t\sqrt{\alpha\beta}}}{e^{2\sqrt{\alpha\beta}} - 1} + \frac{50 \cdot \left(6e^{\sqrt{\alpha\beta}} - 5\right)e^{t\sqrt{\alpha\beta}}}{e^{2\sqrt{\alpha\beta}} - 1}$$

10.4.2 How to solve differential equations numerically?

Some differential equations do not have a closed form solution in terms of elementary functions. For example, the Airy or Stocks equation:

$$\frac{d^2y}{dx^2} - xy$$

Attempting to solve this with Sympy gives:

Jupyter input

```
1  import sympy as sym
2
3  y = sym.Function("y")
4  x = sym.Symbol("x")
5
6  equation = sym.Eq(sym.diff(y(x), x, 2), x * y(x))
7  sym.dsolve(equation, y(x))
```

$$y(x) = C_1 Ai(x) + C_2 Bi(x)$$

which is a linear combination of A_i and B_i which are special functions called the Airy functions of the first and second kind.

Using `scipy.integrate` it is possible to solve this differential equation numerically.

First, define a new variable $u = \frac{dy}{dx}$ so that the second-order differential equation can be expressed as a system of single-order differential equations:

$$\begin{cases} \frac{du}{dx} = & xy \\ \frac{dy}{dx} = & u \end{cases}$$

Now define a python function that returns the right hand side of that system of equations:

Jupyter input

```
1  def diff(state, x):
2      """
3      Returns the value of the derivates for a given set of state values
        ↪ (u, y).
4      """
5      u, y = state
6      return x * y, u
```

You can pass this to `scipy.integrate.odeint` which is a tool that carries out numerical integration of differential equations. Note, that it is incapable of dealing with symbolic variables, thus an initial numeric value of (u, y) is required.

Jupyter input

```
1  import numpy as np
2  import scipy.integrate
3
4  initial_state = (.1, -.5)
5
6  xs = np.linspace(0, 1, 50)
7  states = scipy.integrate.odeint(diff, y0=initial_state, t=xs)
8
```

Here, you make use of numpy to create a collection of x values over which to carry out the numerical integration.

This returns an array of values of states corresponding to (u, y).

```
1    states
```

```
array([[ 0.1       , -0.5       ],
       [ 0.09989617, -0.49795991],
       [ 0.09958578, -0.49592403],
       [ 0.09907053, -0.49389658],
       ...
       [-0.09525243, -0.46835567],
       [-0.10414704, -0.47038996],
       [-0.11327831, -0.47260818],
       [-0.12265169, -0.47501521],
       [-0.13227299, -0.47761605]])
```

Figure 10.1 shows a plot of the above with a comparison to the exact expected values (obtained using the of the first and second kind).

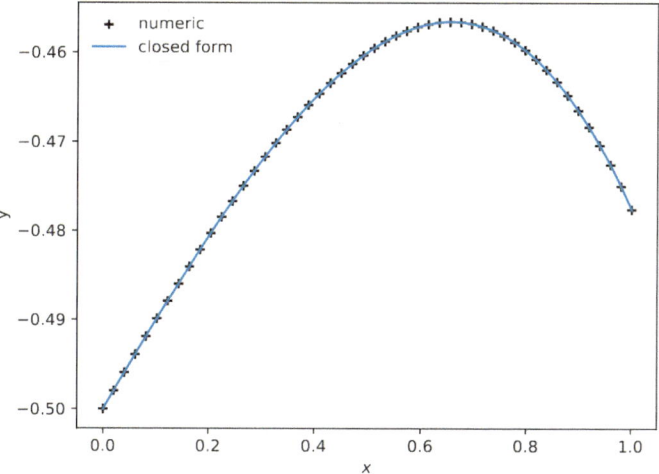

Figure 10.1   Numerical and exact solution to the Stokes differential equation.

Building Tools

CHAPTER 11

Variables, Conditionals and Loops

In the previous chapters you have explored a number of tools that allow you to use mathematical knowledge more efficiently. In this part of the book you will start to gain the knowledge necessary to build such tools.

In this chapter you will cover:

- Creating variables.

- Run code depending on a given condition.

- Repeat code as long as a given condition is met.

- Repeat code over a given iterable.

11.1   TUTORIAL

You will here use a computer to gain some evidence to help tackle the following problem.
Consider the following polynomial:

$$p(n) = n^2 + n + 41$$

1. Verify that $p(n)$ is for $n \in \mathbb{Z}$ up until $n = 20$.

2. What is the smallest value of n for which $p(n)$ is no longer prime?

You will start by defining a function for $p(n)$:

Jupyter input

```
def p(n):
    """
    Return the value of n ^ 2 + n + 41 for a given value of n.
    """
    return n ** 2 + n + 41
```

DOI: 10.1201/9781003451860-11

You will use `sympy` to check if a number is prime.

Jupyter input

```
1  import sympy as sym
2
3  sym.isprime(3)
```

True

Jupyter input

```
1  sym.isprime(4)
```

False

Now to answer the first question you will use a list comprehension to create a list of boolean variables that confirm if $p(n)$ is prime.

> This is similar to what was done in Chapter 7.

Jupyter input

```
1  checks = [sym.isprime(p(n)) for n in range(21)]
2  checks
```

```
[True,
 True,
 True,
 True,
 True,
 True,
 True,
 True,
 True,
 True,
 True,
 True,
 True,
 True,
 True,
 True,
 True,
 True,
 True,
 True,
 True]
```

You can use the `all` tool to check if all the boolean values are true:

Jupyter input

```
1  all(checks)
```

True

> Using list comprehensions is a mathematical way of repeating code but at times it might prove useful to repeat code in a different way using a standard `for` statement.

In that case you can essentially repeat the previous exercise using:

Jupyter input

```
1  checks = []
2  for n in range(21):
3      value = p(n)
4      is_prime = sym.isprime(value)
5      checks.append(is_prime)
6  all(checks)
```

True

The main difference between the two approaches is that you can include multiple lines of indented code to be repeated for every value of n in `range(21)`.

> A `for` loop or a list comprehension should be used when you know how many repetitions are necessary.

To answer the second question you will repeat the code until the value of $p(n)$ is no longer prime.

Jupyter input

```
1  n = 0
2  while sym.isprime(p(n)):
3      n += 1
4  n
```

40

> A `while` loop should be used when you do not know how many times a repetition should be made **but** you know under what conditions it should be made.

Indeed for $n = 40$ you have:

Jupyter input

```
1  p(n)
```

1681

and

Jupyter input

```
1  sym.isprime(p(n))
```

False

sympy can also factor the number:

Jupyter input

```
1  sym.factorint(p(n))
```

{41:2}

Jupyter input

```
1  41 ** 2
```

Indeed:

Jupyter input

```
1  41 ** 2
```

1681

11.2 HOW TO

11.2.1 Define an integer variable

To define an integer variable use the = operator which is the assignment operator. Create the name of the variable, then the assignment operator followed by the integer value.

Usage

```
1  name_of_variable = int_value
```

For example:

Jupyter input

```
1  year = 2020
2  year
```

2020

When choosing a variable name there are some rules to follow:

- No spaces, use _ instead.

- Do not start with a number or other special characters.

There are other important conventions:

- Use explicit names that clearly describe what the variable is. Try not to use i, a unless those refer to specific mathematical variables.

- Do not use CamelCase but use snake_case when combining words. This follows the Python convention called PEP8.

11.2.2 Define a float variable

To define a float variable use the = operator which is the assignment operator. Create the name of the variable, then the assignment operator followed by the real value.

Usage

```
1  name_of_variable = float_value
```

For example:

Jupyter input

```
1  cms_in_an_inch = 2.54
2  cms_in_an_inch
```

2.54

11.2.3 Define a string variable

To define a string variable use the = operator which is the assignment operator. Create the name of the variable, then the assignment operator followed by the string which is a combination of characters between quotation marks.

Usage

```
1   name_of_variable = string_value
```

For example:

Jupyter input

```
1   capital_of_dominica = "roseau"
2   capital_of_dominica
```

`'roseau'`

11.2.4 Define a boolean variable

A boolean variable is one of two things: `True` or `False`. To define a boolean variable you use the = assignment operator. Create the name of the variable then the assignment operator followed by the boolean variable (either `True` or `False`).

Usage

```
1   name_of_variable = boolean_value
```

For example:

Jupyter input

```
1   john_nash_has_a_nobel = True
2   john_nash_has_a_nobel
```

Jupyter input

```
1   True
```

Section 6.2.3 gives an overview of how to create boolean variables from other variables.

11.2.5 Check the type of a variable

You can get the type of a variable using the `type` tool.

Usage

```
1   type(object)
```

Where `object` is any variable.
For example:

Jupyter input

```
1   year = 2020
2   type(year)
```

`int`

Jupyter input

```
1   cms_in_an_inch = 2.54
2   type(cms_in_an_inch)
```

`float`

Jupyter input

```
1   capital_of_dominica = "roseau"
2   type(capital_of_dominica)
```

`str`

If a numeric variable is given with any decimal part (including 0), then it is considered to be a float.

11.2.6   Manipulate numeric variables

Numeric values can be combined to create new numeric variables.

1. Addition, $2 + 2$: `2 + 2`;

2. Subtraction, $3 - 1$: `3 - 1`;

3. Multiplication, 3×5: `3 * 5`;

4. Division, $20/5$: `20 / 5`;

5. Exponentiation, 2^4: `2 ** 4`;

6. Integer remainder, $5 \mod 2$: `5 % 2`; and

7. Combining operations, $\frac{2^3+1}{4}$: `(2 ** 3 + 1) / 4`;

For example:

Jupyter input

```
1  cms_in_an_inch = 2.54
2  average_male_height_in_cms = 170
3  average_male_height_in_inches = average_male_height_in_cms /
   ↪  cms_in_an_inch
4  average_male_height_in_inches
```

```
66.92913385826772
```

This is similar to what what is shown in Section 2.3.6.

Some languages, including Python have a shortcut to manipulate a variable "in place". The following takes the variable money and replaces it by 3 times money:

Jupyter input

```
1  money *= 3
```

This is equivalent to:

Jupyter input

```
1  money = money * 3
```

11.2.7   Include variables in strings

Variables can be used in strings using **string formatting**. There are numerous ways this can be done in Python but the current best practice is to use f-strings.

Usage

```
1  f"{variable}"
```

For example, the following creates a string that uses a random number:

Jupyter input

```
1  import random
2
3  random.seed(0)
4  random_number = random.random()
5  string = f"Here is a random number: {random_number}"
6  string
```

```
'Here is a random number: 0.8444218515250481'
```

11.2.8 Combine collections of boolean variables

Given an iterable of booleans it is possible to check if any or all of them are True using any or all:

> **Usage**
> ```
> 1   all(iterable)
> ```

> **Usage**
> ```
> 1   any(iterable)
> ```

For example:

> **Jupyter input**
> ```
> 1   iterable = (True, True, False, True, True)
> 2   all(iterable)
> ```

False

> **Jupyter input**
> ```
> 1   any(iterable)
> ```

True

11.2.9 Run code **if** a condition holds

An important part of giving instructions to a computer is to specify when to do different things. This is done using what is called an if statement. Following an if, a boolean variable is expected, if that boolean is True then the indented code that follows is run. Otherwise it is not.

Usage

```
1  if boolean:
2      code to run if boolean is true
3  else:
4      code to run if boolean is false
5  code to run after either of two previous code blocks are run.
```

An `else` statement is not always necessary. Specifically when combined with functions as seen in Chapter 7 the `else` is often not needed.

For example, the following code selects a random integer between 0 and 100 and then returns a different string depending on what the number was.

Jupyter input

```
1  import random
2
3  random.seed(0)
4  random_number = random.randint(0, 100)
5  is_even = random_number % 2 == 0
6  if is_even:
7      message = f"The random number ({random_number}) is even."
8  else:
9      message = f"The random number ({random_number}) is odd."
10 message
```

```
'The random number (49) is odd.'
```

11.2.10 Repeat code **for** a given set of variables

Given an iterable, it is possible to repeat some code for every item in the iterable. This is done using what is called a `for` loop. Following the `for` a placeholder variable is given, then followed by the `in` keyword and the iterable. After that the indented code that will be repeated for every value of the iterable.

Usage

```
1  for placeholder_variable in iterable:
2      code to repeat
```

For example, the following will print a message for every given value in the iterable:

```
   Jupyter input

1  iterable = ("Dog", 3, 2, -1.0)
2  for item in iterable:
3      type_of_variable = type(item)
4      message = f"The variable {item} has type {type_of_variable}"
5      print(message)
```

```
The variable Dog has type <class 'str'>
The variable 3 has type <class 'int'>
The variable 2 has type <class 'int'>
The variable -1.0 has type <class 'float'>
```

for loops are a common tool across most programming languages. They are similar to the list comprehensions seen in Section 7.2.5.

- List comprehensions should be specifically used when the goal is to create a collection of items.

- Traditional for loops should be used when the code to run for every iteration is more complex.

A common use case of for loops is to combine them with a range statement to repeat code a known number of items.

11.2.11 Repeat code **while** a given condition holds

To repeat code while a condition holds a while loop should be used. Similarly to the if statement, following a while, a boolean variable is expected, if that boolean is True, then the indented code that follows is repeated. After it is run, the boolean is checked once more. When the boolean is False the indented code is skipped.

```
   Usage

1  while boolean:
2      code to repeat before checking boolean once more
3  code to run once boolean is False
```

Here is some code that repeatedly selects a random integer until that number is even.

Jupyter input

```
1  import random
2
3  random.seed(4)
4  selected_integer = random.randint(0, 10)
5  number_of_selections = 1
6  while selected_integer % 2 == 1:
7      selected_integer = random.randint(0, 10)
8      number_of_selections += 1
9  number_of_selections
```

2

11.2.12 Iterate over pairs of items from two iterables

To create a new iterable of pairs of items from two separate iterables use `zip`:

Usage

```
1  zip(iterable_1, iterable_2)
```

For example:

Jupyter input

```
1  basket = ("Carrots", "Potatoes", "Strawberries", "Juice", "Ice cream")
2  prices = (4, 2, 6, 3, 10)
3  pairs = [(item, price) for item, price in zip(basket, prices)]
4  pairs
```

```
[('Carrots', 4),
 ('Potatoes', 2),
 ('Strawberries', 6),
 ('Juice', 3),
 ('Ice cream', 10)]
```

11.2.13 Iterate over and index items from an iterable

To iterate over items from an iterable and keep track of their index use `enumerate`:

Usage

```
1  enumerate(iterable)
```

For example:

```
1  basket = ("Carrots", "Potatoes", "Strawberries", "Juice", "Ice cream")
2  indices_and_items = [(count, item) for count, item in
   ↪  enumerate(basket)]
3  indices_and_items
```

```
[(0, 'Carrots'),
 (1, 'Potatoes'),
 (2, 'Strawberries'),
 (3, 'Juice'),
 (4, 'Ice cream')]
```

11.3 EXERCISES

1. Using a `for` loop print the types of the variables in each of the following iterables:

   (a) `iterable = (1, 2, 3, 4)`

   (b) `iterable = (1, 2.0, 3, 4.0)`

   (c) `iterable = (1, "dog", 0, 3, 4.0)`

2. Consider the following polynomial:

$$3n^3 - 183n^2 + 3318n - 18757$$

   (a) Use the `sympy.isprime` function to find the lowest positive integer value of n for which the absolute value of that polynomial is not prime?

   (b) How many **unique** primes up until the first non-value are there? (Hint: the `set` tool might prove useful here.)

3. Check the following identity for each value of $n \in \{0, 10, 100, 2000\}$:

$$\sum_{i=0}^{n} i = \frac{n(n+1)}{2}$$

4. Check the following identity for all positive integer values of n less than 5000:

$$\sum_{i=0}^{n} i^2 = \frac{n(n+1)(2n+1)}{6}$$

5. Repeat the experiment of selecting a random integer between 0 and 10 until it is even 1000 times. What is the average number of times taken to select an even number?

11.4 FURTHER INFORMATION

11.4.1 Why can I not only use a `while` loop?

The `for` loop allows you to iterate over any selection of objects. Some languages do not have a generic `for` loop. In some cases it is only possible to iterate over a set of integers (similar to the `for i in range(n)` pattern) or to only use a `while` loop.

Because of this, it is often the case that you will see code that uses `while` loops instead of `for` loops. For example:

Jupyter input

```
1   seasons = ("Winter", "Spring", "Summer", "Autumn")
2
3   number_of_seasons = len(seasons)
4   i = 0
5   while i < number_of_seasons:
6       season = seasons[i]
7       print(season)
8       i += 1
```

```
Winter
Spring
Summer
Autumn
```

The above code is equivalent to:

Jupyter input

```
1   seasons = ("Winter", "Spring", "Summer", "Autumn")
2   for season in seasons:
3       print(season)
```

```
Winter
Spring
Summer
Autumn
```

While it is possible to use a `while` loop instead of a `for` loop there are no advantages to doing that and in fact only disadvantages:

- Using the `while` loop requires iterating over the iterable twice: the first time when counting the length of it using `len` and the second time during the `while` statement itself.

- There is more potential for error in the code: it would not be unlikely to have an off by one error in the boolean condition.

- It is less readable.

The following is a good guideline:

- Use a `for` loop when you know what you are iterating over.

- Use a `while` loop when only know a specific condition under which you should iterate.

11.4.2 Why should I not check if a boolean is equal to `True` or `False`

It is possible to create a boolean by comparing another boolean to `True` or `False` for example:

> **Jupyter input**
> ```
> 1  boolean = False
> 2  boolean == True
> ```

False

Thus when using `if` or `while` statements you might sometimes see things like the following:

> **Jupyter input**
> ```
> 1  import random
> 2
> 3  random.seed(4)
> 4  selected_integer = random.randint(0, 10)
> 5  number_of_selections = 1
> 6  while (selected_integer % 2 == 1) == True:
> 7      selected_integer = random.randint(0, 10)
> 8      number_of_selections += 1
> 9  number_of_selections
> ```

2

or:

> **Jupyter input**
> ```
> 1  random.seed(4)
> 2  selected_integer = random.randint(0, 10)
> 3  number_of_selections = 1
> 4  while (selected_integer % 2 == 1):
> 5      selected_integer = random.randint(0, 10)
> 6      number_of_selections += 1
> 7  number_of_selections
> ```

2

However, this is not best practice. A better approach is to use `is` instead of `==`:

```
Jupyter input
1   import random
2
3   random.seed(4)
4   selected_integer = random.randint(0, 10)
5   number_of_selections = 1
6   while (selected_integer % 2 == 1) is True:
7       selected_integer = random.randint(0, 10)
8       number_of_selections += 1
9   number_of_selections
```

2

This is due to the fact that when using `==` variables that are not booleans will be converted to booleans and this might not be the expected behaviour.

For example:

```
Jupyter input
1   number = 0
2   number == False
```

True

however:

```
Jupyter input
1   number is False
```

False

Functions and Data Structures

In the previous chapters you have explored a number of tools that allow you to use your mathematical knowledge more efficiently. In this chapter you continue to gain the knowledge necessary to build these tools covering the following topics:

> In this chapter you will cover:
>
> - Defining and using functions.
>
> - Defining and using various data structures.

12.1 TUTORIAL

Similarly to Chapter 11, you will use a computer to gain numerical evidence for a problem. Consider the following sequence:

$$\begin{cases} a_0 = 0, \\ a_1 = 1, \\ a_n = a_{n-1} + a_{n-2}, n \geq 2 \end{cases}$$

Verify that the following identity holds for $n \leq 500$:

$$\sum_{i=0}^{n} a_i = a_{n+2} - 1$$

You will start by defining a function for $a(n)$:

Jupyter input

```
1   import functools
2
3
4   @functools.lru_cache()
5   def get_fibonacci(n):
6       """
7       A function to give the nth Fibonacci number using the recursive
8       definition.
9
10      Note that this also uses a cache.
11
12      Parameters
13      ----------
14      n: int
15          The index of the Fibonacci number
16
17      Returns
18      -------
19      int
20          The nth Fibonacci number
21      """
22      if n == 0:
23          return 0
24      if n == 1:
25          return 1
26      return get_fibonacci(n - 1) + get_fibonacci(n - 2)
```

This uses caching in the function definition with `lru_cache`. This is not necessary but makes the code more efficient. Caching is covered in Section 8.4.2.

You will print the first ten numbers to ensure everything is working correctly:

Jupyter input

```
1   for n in range(10):
2       print(get_fibonacci(n))
```

```
0
1
1
2
3
5
8
13
```

21
34

Now write a function that returns a boolean: True if the equation holds for a given value of n, False otherwise.

Jupyter input

```python
def check_theorem(n):
    """
    A function that generate the lhs and rhs of the
    following relationship:

    \sum_{i=0}^n a_i = a_{n + 2} - 1

    Where `a_i` is the i-th Fibonacci number.

    It checks if the relationship holds.

    Parameters
    ----------
    n: int
        The index n for which the theorem is to be verified.

    Returns
    -------
    bool
        Whether or not the theorem holds for a given n.
    """
    sum_of_fibonacci = sum(get_fibonacci(i) for i in range(n + 1))
    return sum_of_fibonacci == get_fibonacci(n + 2) - 1
```

Generate checks for $n \leq 500$:

Jupyter input

```python
checks = [check_theorem(n) for n in range(501)]
checks
```

```
[True,
 True,
 True,
 ...
 True,
 True,
 True]
```

Confirm that all the booleans in checks are True:

```
1  all(checks)
```

True

12.2   HOW TO

Two important data structures have already been seen in previous chapters:

- Tuples: Section 6.2.1.
- Lists: Section 7.2.1.

12.2.1   Define a function

See Section 7.2.2.

12.2.2   Write a docstring

A docstring is an attribute of a function that describes what it is. This can describe what it does, how it does it and/or why it does it. Here is how to write a docstring for a function that takes variables and returns a value.

Usage

```
1   def name(parameter1, parameter2, ...):
2       """
3       <A description of what the function is.>
4
5       Parameters
6       ----------
7       parameter1 : <type of parameter1>
8           <description of parameter1>
9       parameter2 : <type of parameter2>
10          <description of parameter2>
11      ...
12
13      Returns
14      -------
15      <type of what the function returns>
16          <description of what the function returns>
17
18      """
19      INDENTED BLOCK OF CODE
20      return output
```

For example, here is how to write a function that returns x^3 for a given x:

```
   Jupyter input
1  def x_cubed(x):
2      """
3      Calculates and returns the cube of x. Does this by using Python
4      exponentiation.
5
6      Parameters
7      ----------
8      x : float
9          The value of x to be raised to the power 3
10
11     Returns
12     -------
13     float
14         The cube.
15     """
16     return x ** 3
```

12.2.3 Create a tuple

See Section 6.2.1.

12.2.4 Create a list

See Section 7.2.1.

12.2.5 Create a list using a list comprehension

See Section 7.2.5.

12.2.6 Combine lists

Given two lists it is possible to combine them to create a new list using the + operator:

```
   Usage
1  first_list + other_list
```

Here is an example of creating a single list from two separate lists:

Jupyter input

```
1  first_list = [1, 2, 3]
2  other_list = [5, 6, 100]
3  combined_list = first_list + other_list
4  combined_list
```

```
[1, 2, 3, 5, 6, 100]
```

12.2.7 Append an element to a list

Appending an element to a list is done using the **append** method.

Usage

```
1  a_list.append(element)
```

Here is an example where you append a new string to a list of strings:

Jupyter input

```
1  names = ["Vince", "Zoe", "Julien", "Kaitlynn"]
2  names.append("Riggins")
3  names
```

```
['Vince', 'Zoe', 'Julien', 'Kaitlynn', 'Riggins']
```

> It is not possible to do this with a **tuple** as a **tuple** is **immutable**. See Section 7.4.1 for more information on the difference between a list and a tuple.

12.2.8 Remove an element from a list

To remove a given element from a list use the **remove** method.

Usage

```
1  a_list.remove(element)
```

Here is an example removing a number from a list of numbers:

Jupyter input

```
1   numbers = [1, 94, 23, 202, 5]
2   numbers.remove(23)
3   numbers
```

```
[1, 94, 202, 5]
```

It is not possible to remove an element from a `tuple` as a `tuple` is immutable. See Section 7.4.1 for more information on the difference between a list and a tuple.

12.2.9   Sort a list

To sort a list use the `sort` method.

Usage

```
1   a_list.sort()
```

Here is an example:

Jupyter input

```
1   names = ["Vince", "Zoe", "Kaitlynn", "Julien"]
2   names.sort()
3   names
```

```
['Julien', 'Kaitlynn', 'Vince', 'Zoe']
```

To sort a list in reverse order use the `sort` method with the `reverse=True` parameter.

Jupyter input

```
1   names.sort(reverse=True)
2   names
```

```
['Zoe', 'Vince', 'Kaitlynn', 'Julien']
```

It is not possible to sort a `tuple` as a `tuple` is immutable. See Section 7.4.1 for more information on the difference between a list and a tuple.

12.2.10   Create a sorted list from an iterable

To create a sorted list from an iterable use the `sorted` function.

Usage

```
1  sorted(iterable)
```

Here is an example:

Jupyter input

```
1  tuple_of_numbers = (20, 50, 10, 6, 1, 50, 105)
2  sorted(tuple_of_numbers)
```

```
[1, 6, 10, 20, 50, 50, 105]
```

12.2.11 Access an element of an iterable

See Section 6.2.2.

12.2.12 Find the index of an element in an iterable

To identify the position of an element in an iterable use the `index` method.

Usage

```
1  iterable.index(element)
```

Here is an example:

Jupyter input

```
1  numbers = [1, 94, 23, 202, 5]
2  numbers.index(23)
```

```
2
```

> Recall that python uses 0-based indexing. The first element in an iterable has index 0.

12.2.13 Access an element of an iterable using negative indexing

It is possible to access an element of an iterable by counting from the end of the iterable using negative indexing.

Usage

```
1   iterable[-index_from_end]
```

Here is an example showing how to access the penultimate element in a tuple:

Jupyter input

```
1   basket = ("Carrots", "Potatoes", "Strawberries", "Juice", "Ice cream")
2   basket[-2]
```

`'Juice'`

12.2.14   Slice an iterable

To create a new iterable from an iterable use [] and specify a start (inclusive) and end (exclusive) pair of indices.

Usage

```
1   iterable[include_start_index: exclusive_end_index]
```

For example:

Jupyter input

```
1   basket = ("Carrots", "Potatoes", "Strawberries", "Juice", "Ice cream")
2   basket[2: 5]
```

`('Strawberries', 'Juice', 'Ice cream')`

12.2.15   Find the number of elements in an iterable

To count the number of elements in an iterable use `len`:

Usage

```
1   len(iterable)
```

For example:

```
1  basket = ("Carrots", "Potatoes", "Strawberries", "Juice", "Ice cream")
2  len(basket)
```

5

12.2.16 Create a set

A set is a collection of distinct objects. This can be created in Python using the `set` command on any iterable. If there are non-distinct objects in the iterable, then this is an efficient way to remove duplicates.

Usage

```
1  set(iterable)
```

Here is an example of creating a set:

Jupyter input

```
1  iterable = (1, 1, 3, 4, 4, 3, 2, 1, 10)
2  unique_values = set(iterable)
3  unique_values
```

```
{1, 2, 3, 4, 10}
```

12.2.17 Do set operations

Set operations between two sets can be done using Python:

- $S_1 \cup S_2$: set_1 | set_2

- $S_1 \cap S_2$: set_1 & set_2

- $S_1 \setminus S_2$: set_1 - set_2

- $S_1 \subseteq S_2$ (checking if S_1 is a subset of S_2): set_1 <= set_2

Here are some examples of carrying out the above:

Jupyter input

```
1  set_1 = set((1, 2, 3, 4, 5))
2  set_2 = set((4, 5, 6, 7, 8, 9))
3
4  set_1 | set_2
```

{1, 2, 3, 4, 5, 6, 7, 8, 9}

Jupyter input

```
1  set_1 & set_2
```

{4, 5}

Jupyter input

```
1  set_1 - set_2
```

{1, 2, 3}

Jupyter input

```
1  set_1 <= set_2
```

False

12.2.18 Create hash tables

Lists and tuples allow us to immediately recover a value given its position. Hash tables allow us to create arbitrary `key value` pairs so that given any `key` you can immediately recover the value. This is called a dictionary in Python and is created using {} which takes a collection of `key: value` pairs.

Usage

```
1  {key_1: value, key_2: value, ...}
```

For example, the following dictionary maps pet names to their ages:

Jupyter input

```
1   ages = {"Riggins": 4, "Chick": 7, "Duck": 7}
2   ages
```

```
{'Riggins': 4, 'Chick': 7, 'Duck': 7}
```

To recover a value pass the key to the dictionary using [].
For example:

Jupyter input

```
1   ages["Riggins"]
```

```
4
```

> If a key is used to recover the value with [] but the key is not in the dictionary, then an error will be raised.

12.2.19   Access element in a hash table

As described in Section12.2.18 to access the value of a key in a hash table, use [].

Usage

```
1   dictionary[key]
```

It is also possible to use the get method. The get method can also be passed the value of a default variable to return when the key is not in the hash table:

Jupyter input

```
1   dictionary.get(key, default)
```

For example:

Jupyter input

```
1   ages = {"Riggins": 4, "Chick": 7, "Duck": 7}
2   ages.get("Vince", -1)
```

```
-1
```

12.2.20 Iterate over keys in a hash table

To iterate over the keys in a hash table, use the `keys()` method:

> **Usage**
>
> ```
> 1  dictionary.keys()
> ```

For example:

> **Jupyter input**
>
> ```
> 1  ages = {"Riggins": 4, "Chick": 7, "Duck": 7}
> 2  ages.keys()
> ```

```
dict_keys(['Riggins', 'Chick', 'Duck'])
```

12.2.21 Iterate over values in a hash table

To iterate over the values in a hash table, use the `values()` method:

> **Usage**
>
> ```
> 1  dictionary.values()
> ```

For example:

> **Jupyter input**
>
> ```
> 1  ages = {"Riggins": 4, "Chick": 7, "Duck": 7}
> 2  ages.values()
> ```

> **Jupyter input**
>
> ```
> 1  dict_values([4, 7, 7])
> ```

12.2.22 Iterate over pairs of keys and value in a hash table

To iterate over pairs of keys and values in a hash table, use the `items()` method:

> **Usage**
>
> 1   `dictionary.items()`

For example:

> **Jupyter input**
>
> 1   `ages = {"Riggins": 4, "Chick": 7, "Duck": 7}`
> 2   `ages.items()`

```
dict_items([('Riggins', 4), ('Chick', 7), ('Duck', 7)])
```

12.3 EXERCISES

1. Write a function that generates $(n!)$.

2. Write a function that generates the nth triangular numbers defined by:

$$T_n = \frac{n(n+1)}{2}$$

3. Verify the following that the following identify holds for positive integer values $n \leq 500$:

$$\sum_{i=0}^{n} T_i = \frac{n(n+1)(n+2)}{6}$$

4. Consider the **Monty Hall problem** [11]:

   "Suppose you're on a game show, and you're given the choice of three doors: Behind one door is a car; behind the others, goats. You pick a door, say No. 1, and the host, who knows what's behind the doors, opens another door, say No. 3, which has a goat. He then says to you, 'Do you want to pick door No. 2?'

   Is it to your advantage to switch your choice?"

   (a) Write a function that simulates the play of the game when you "stick" with the initial choice. You might find `random.shuffle` and poping a list helpful.

   (b) Write a function that simulates the play of the game when you "change" your choice. You might find removing from a list helpful.

   (c) Repeat the play of the game using both those functions and compare the probability of winning.

12.4 FURTHER INFORMATION

12.4.1 What formats can be used to write a docstring?

The format used to write a docstring described in Section 12.2.2. is the one specified by the Numpy project.

Amongst other things you can see how to specify further functionality:

- How to indicate if a parameter is optional.

- How to specify what types of errors might be raised by a function.

- How to specify when a function is a generator.

There are 2 other common specifications:

- Google's Python Style Guide.

- The Sphinx Python Style Guide.

12.4.2 Are there tools available to assist with writing docstrings?

The `darglint` library can be used to check if docstrings match a given format.

12.4.3 Apart from removing duplicates and set operations what are the advantages of using `set`?

One valuable uses of `set` is to efficiently identify if an element is in a given iterable or not:

```
Jupyter input

1   numbers = list(range(100000))
2   %timeit 100000 in numbers
```

474 µs ± 2.51 µs per loop (mean ± std. dev. of 7 runs, 1,000 loops each)

```
Jupyter input

1   numbers = set(range(100000))
2   %timeit 100000 in numbers
```

15.2 ns ± 0.121 ns per loop (mean ± std. dev. of 7 runs, 100,000,000 loops each)

CHAPTER 13

Object-Oriented Programming

In the first part of this book you covered a number of tools that allow you to carry out mathematical techniques. One example of this is the `sympy.Symbol` object that creates a symbolic variable. In this chapter you will see how to define similar mathematical objects.

In this chapter you will cover:

- Creating objects.

- Giving objects attributes.

- Defining methods on objects.

- Inheriting new objects from others.

13.1 TUTORIAL

You will write some code to create and manipulate expressions. With `sympy` this is not necessary as all functionality required is available within `sympy`; however, this will be a good exercise in understanding how to build such functionality.

Consider the following quadratics:

$$f(x) = 5x^2 + 2x - 7$$
$$g(x) = -4x^2 - 3x + 12$$
$$h(x) = f(x) + g(x)$$

Without using `sympy`, obtain the roots for all the quadratics.

Start by defining an object to represent a quadratic. This is called a class.

DOI: 10.1201/9781003451860-13

Jupyter input

```python
import math


class QuadraticExpression:
    """A class for a quadratic expression"""

    def __init__(self, a, b, c):
        self.a = a
        self.b = b
        self.c = c
        self.discriminant = self.b ** 2 - 4 * self.a * self.c

    def get_roots(self):
        """
        Return the real valued roots of the quadratic expression

        Returns
        -------
        array
            The roots of the quadratic
        """
        if self.discriminant >= 0:
            x1 = -(self.b + math.sqrt(self.discriminant)) / (2 * self.a)
            x2 = -(self.b - math.sqrt(self.discriminant)) / (2 * self.a)
            return x1, x2
        return ()

    def __add__(self, other):
        """A magic method: let's us have addition between
        ↪  expressions"""
        return QuadraticExpression(self.a + other.a, self.b + other.b,
        ↪  self.c + other.c)

    def __repr__(self):
        """A magic method: changes the default way an instance is
        ↪  displayed"""
        return f"Quadratic expression: {self.a} x ^ 2 + {self.b} x +
        ↪  {self.c}"
```

Four functions were created with this class:

- __init__: as this is surrounded by __ (two underscores) this is a magic function that is run when we create an instance of our class.

- get_roots: this returns the two real valued roots if the is positive.

- __add__: a magic function that is run when the + operator is used.

- __repr__: a magic function that gives the string representation of the instance.

Now use this class to solve the specified problem. First create instances of the class that correspond to f and g. This is using the __init__ function in the background.

Jupyter input

```
1  f = QuadraticExpression(a=5, b=2, c=-7)
2  g = QuadraticExpression(a=-4, b=-3, c=12)
```

You can now take a look at both of these instances. This is using the __repr__ function in the background:

Jupyter input

```
1  f
```

Quadratic expression: 5 x ^ 2 + 2 x + -7

Jupyter input

```
1  g
```

Quadratic expression: -4 x ^ 2 + -3 x + 12

Now you are going to create $h(x) = f(x) + g(x)$. This is using the __add__ function in the background:

Jupyter input

```
1  h = f + g
2  h
```

Quadratic expression: 1 x ^ 2 + -1 x + 5

You can now iterate over the quadratics and find the roots. This is using the get_roots function in the background:

Jupyter input

```
1  roots = [quadratic.get_roots() for quadratic in (f, g, h)]
2  roots
```

[(-1.4, 1.0), (1.3971808598447282, -2.1471808598447284), ()]

Note that f and g have real valued roots but h does not. You can check the value of the discriminant of h:

Jupyter input

```
1  h.discriminant
```

-19

You are going to create a new class from QuadraticExpression replacing the get_roots function with a new one that can handle imaginary roots and update the __add__ function to return an instance of the new class.

Jupyter input

```
1   class QuadraticExpressionWithAllRoots(QuadraticExpression):
2       """
3       A class for a quadratic expression that can return imaginary roots
4
5       The `get_roots` function returns two tuples of the form (re, im)
         ↪  where re is
6       the real part and im is the imaginary part.
7       """
8
9       def get_roots(self):
10          """
11          Return the real valued roots of the quadratic expression
12
13          Returns
14          -------
15          array
16              The roots of the quadratic
17          """
18          if self.discriminant >= 0:
19              x1 = -(self.b + math.sqrt(self.discriminant)) / (2 * self.a)
20              x2 = -(self.b - math.sqrt(self.discriminant)) / (2 * self.a)
21              return (x1, 0), (x2, 0)
22
23          real_part = self.b / (2 * self.a)
```

```
24          im1 = math.sqrt(-self.discriminant) / (2 * self.a)
25          im2 = -math.sqrt(-self.discriminant) / (2 * self.a)
26          return ((real_part, im1), (real_part, im2))
27
28      def __add__(self, other):
29          """A special method: let's us have addition between
            ↪ expressions"""
30          return QuadraticExpressionWithAllRoots(
31              self.a + other.a, self.b + other.b, self.c + other.c
32          )
```

Now define the quadratics once again but using this new class:

Jupyter input

```
1   f = QuadraticExpressionWithAllRoots(a=5, b=2, c=-7)
2   g = QuadraticExpressionWithAllRoots(a=-4, b=-3, c=12)
3   h = f + g
```

Jupyter input

```
1   f
```

Quadratic expression: 5 x ^ 2 + 2 x + -7

Jupyter input

```
1   g
```

Quadratic expression: -4 x ^ 2 + -3 x + 12

Jupyter input

```
1   h
```

Jupyter input

```
1   Quadratic expression: 1 x ^ 2 + -1 x + 5
```

There is no need to redefine `__init__`, or `__repr__` as the new class inherits these from `QuadraticExpression` due to this statement:

Jupyter input

```
1  class QuadraticExpressionWithAllRoots(QuadraticExpression):
```

You can now get all the roots for the quadratics:

Jupyter input

```
1  roots = [quadratic.get_roots() for quadratic in (f, g, h)]
2  roots
```

```
[((-1.4, 0), (1.0, 0)),
 ((1.3971808598447282, 0), (-2.1471808598447284, 0)),
 ((-0.5, 2.179449471770337), (-0.5, -2.179449471770337))]
```

13.2   HOW TO

13.2.1   Define a class

Define a class using the `class` keyword:

Usage

```
1  class Name:
2      """
3      A docstring between triple quotation to describe what the class
         ↪  represents
4      """
5      INDENTED BLOCKS OF CODE
```

For example to create a class for a country:

Jupyter input

```
1  class Country:
2      """
3      A class to represent a country
4      """
```

13.2.2   Create an instance of the class

Once a class is defined call it using ():

> **Usage**
>
> ```
> 1   Name()
> ```

For example:

> **Jupyter input**
>
> ```
> 1   first_country = Country()
> 2   first_country
> ```

```
<__main__.Country at 0x7f22a8f76e00>
```

> **Jupyter input**
>
> ```
> 1   second_country = Country()
> 2   second_country
> ```

```
<__main__.Country at 0x7f22a8f76e30>
```

The at `<__main__Country at 0x7f22a8f76e30>` is a pointer to the location of the instance in memory. If you re-run the code that location will change.

13.2.3   Create an attribute

Attributes are variables that belong to instances of classes. They can be created and accessed using `.name_of_variable`.

For example, the following creates the attributes `name` and `amount_of_magic`:

> **Jupyter input**
>
> ```
> 1   first_country.name = "narnia"
> 2   first_country.amount_of_magic = 500
> ```

You can access them:

> **Jupyter input**
>
> ```
> 1   first_country.name
> ```

```
'Narnia'
```

Jupyter input

```
1   first_country.amount_of_magic
```

500

You can manipulate them in place:

Jupyter input

```
1   first_country.amount_of_magic += 100
2   first_country.amount_of_magic
```

600

13.2.4   Create and call a method

Methods are functions that belong to classes. Define a function using the def keyword (short for define). The first variable of a method is always the specific instance that will call the method (it is passed implicitly).

Usage

```
1   class Name:
2       """
3       A docstring between triple quotation to describe what the class
        ↪ represents
4       """
5       def name(self, parameter1, parameter2, ...):
6           """
7           <A description of what the method is.>
8
9           Parameters
10          ----------
11          parameter1 : <type of parameter1>
12              <description of parameter1>
13          parameter2 : <type of parameter2>
14              <description of parameter2>
15          ...
16
17          Returns
18          -------
19          <type of what the function returns>
20              <description of what the function returns>
21
22          """
```

```
23          INDENTED BLOCK OF CODE
24          return output
```

For example, let us create a class for a country that has the ability to "spend" magic:

Jupyter input

```python
1  class Country:
2      """
3      A class to represent a country
4      """
5
6      def spend_magic(self, amount_spent):
7          """
8          Updates the magic attribute by subtracting amount_spent
9
10         Parameters
11         ----------
12         amount_spent : float
13             The amount of mana used.
14         """
15         self.amount_of_magic -= amount_spent
```

Now use it:

Jupyter input

```python
1  first_country = Country()
2  first_country.name = "Narnia"
3  first_country.amount_of_magic = 500
4  first_country.spend_magic(amount_spent=100)
5  first_country.amount_of_magic
```

```
400
```

Even though the method is defined as taking two variables as inputs: `self` and `amount_spent` you only have to explicitly pass it `amount_spent`. The first variable in a method definition always corresponds to the instance on which the method exists.

13.2.5 How to create and call magic methods

Some methods can be called in certain situations:

- When creating an instance.
- When wanting to display an instance.

These are referred to as dunder methods as they are all in between two underscores: __.

The method that is called when an instance is created is called __init__ (for initialised). For example:

Usage

```
1  class Country:
2      """
3      A class to represent a country
4      """
5
6      def __init__(self, name, amount_of_magic):
7          self.name = name
8          self.amount_of_magic = amount_of_magic
```

Now instead of creating an instance and then creating the attributes you can do those two things at the same time, by passing the variables to the class itself (which in turn passes them to the __init__ method):

Jupyter input

```
1  first_country = Country("Narnia", 500)
2  first_country.name
```

'Narnia'

Jupyter input

```
1  first_country.amount_of_magic
```

500

The method that returns a representation of an instance is __repr__. For example:

Jupyter input

```
1  class Country:
2      """
3      A class to represent a country
4      """
5
6      def __init__(self, name, amount_of_magic):
7          self.name = name
8          self.amount_of_magic = amount_of_magic
9
```

```
10    def __repr__(self):
11        """Returns a string representation of the instance"""
12        return f"{self.name} with {self.amount_of_magic} magic."
```

Jupyter input

```
1    first_country = Country("Narnia", 500)
2    first_country
```

```
Narnia with 500 magic.
```

There are numerous other magic methods, such as the __add__ one used in the tutorial of this chapter.

13.2.6 Use inheritance

Inheritance is a tool that allows you to create one class based on another. This is done by passing the Old class to the New class.

Usage

```
1    class New(Old)
```

For example, create a class of MuggleCountry that overwrites the __init__ method but keeps the __rep__ method of the Country class written in Section 13.2.5:

Jupyter input

```
1    class MuggleCountry(Country):
2        """
3        A class to represent a country with no magic. It only requires the
         ↪  name on
4        initialisation.
5        """
6
7        def __init__(self, name):
8            self.name = name
9            self.amount_of_magic = 0
```

This has replaced the __init__ method but the __repr__ method is the same:

> **Jupyter input**
>
> ```
> 1  other_country = MuggleCountry("Wales")
> 2  other_country
> ```

Wales `with` 0 magic.

13.3 EXERCISES

1. Use the class created in Section 13.1 to find the roots of the following quadratics:

    (a) $f(x) = -4x^2 + x + 6$
    (b) $g(x) = 3x^2 - 6$
    (c) $h(x) = f(x) + g(x)$

2. Write a class for a and use it to find the roots of the following expressions:

    (a) $f(x) = 2x + 6$
    (b) $g(x) = 3x - 6$
    (c) $h(x) = f(x) + g(x)$

3. If rain drops were to fall randomly on a square of side length $2r$, the probability of the drops landing in an inscribed circle of radius r would be given by:

$$P = \frac{\text{Area of circle}}{\text{Area of square}} = \frac{\pi r^2}{4r^2} = \frac{\pi}{4}$$

Thus, if you can approximate P, then you can approximate π as $4P$. In this question you will write code to approximate P using the random library.

First create the following class:

> **Jupyter input**
>
> ```
> 1   class Drop:
> 2       """
> 3       A class used to represent a random rain drop falling on a
>     ↪    square of
> 4       length r.
> 5       """
> 6
> 7       def __init__(self, r=1):
> 8           self.x = (0.5 - random.random()) * 2 * r
> 9           self.y = (0.5 - random.random()) * 2 * r
> 10          self.in_circle = (self.y) ** 2 + (self.x) ** 2 <= r ** 2
> ```

Note that the above uses the following equation for a circle centred at $(0,0)$ of radius r:

$$x^2 + y^2 \leq r^2$$

To approximate P create $N = 1000$ instances of Drops and count the number of those that are in the circle. Use this to approximate π.

4. In a similar fashion to question 3, approximate the integral $\int_0^1 1 - x^2 \, dx$. Recall that the integral corresponds to the area under a curve.

13.4   FURTHER INFORMATION

13.4.1   How to pronounce the double underscore?

The double underscore used in magic methods like __init__ or __repr__ is pronounced "dunder".

13.4.2   What is the self variable for?

In methods the first variable is used to refer to the instance of a given class. It is conventional to use self.

As an example consider this class:

Jupyter input

```python
class PetDog:
    """
    A class for a Pet.

    Has two methods:
        - `bark` which returns "Woof" as a string.
        - `give_toy` which gives a toy to the dog in question. This
    ↪    updates the
          `toys` attribute.
    """

    def __init__(self):
        self.toys = []

    def bark(self):
        """
        Returns the string Woof.
        """
        return "Woof"

    def give_toy(self, toy):
        """
        Updates the instances toys list.
        """
        self.toys.append(toy)
```

If you now create two dogs:

Jupyter input

```
1  auraya = PetDog()
2  riggins = PetDog()
```

Both have no toys:

Jupyter input

```
1  auraya.toys
```

[]

Jupyter input

```
1  riggins.toys
```

[]

Now when you want to give `riggins` a toy you need to specify which of those two empty lists to update:

Jupyter input

```
1  riggins.give_toy("ball")
2  riggins.toys
```

['ball']

However, `auraya` still has no toys:

Jupyter input

```
1  auraya.toys
```

[]

When running `riggins.give_toy("ball")`, internally the `give_toy` method is taking `self` to be `riggins` and so the line `self.toys.append(toy)` in fact is running as `riggins.toys.append(toy)`.

The variable name `self` is a convention and not a functional requirement. If you modify it (for example by using inheritance as explained in Section 13.2.6):

```
     Jupyter input

1    class OtherPetDog(PetDog):
2        """
3        A class for a Pet.
4
5        Has two methods:
6            - `bark` which returns "Woof" as a string.
7            - `give_toy` which gives a toy to the dog in question. This
         ↪   updates the
8              `toys` attribute.
9        """
10
11       def give_toy(the_dog_in_question, toy):
12           """
13           Updates the instances toys list.
14           """
15           the_dog_in_question.toys.append(toy)
```

Then you get the same outcome:

```
     Jupyter input

1    riggins = OtherPetDog()
2    riggins.toys
```

[]

```
     Jupyter input

1    riggins.give_toy("ball")
2    riggins.toys
```

['ball']

Indeed the line the_dog_in_question.toys.append(toy) is run as riggins.toys.append(toy). You should however use self as it is convention and helps with readability of your code.

13.4.3 Why use CamelCase for classes but snake_case for functions?

This is specified by the Python convention which is called PEP8 [12].

These conventions are important as it helps with readability of code.

13.4.4 What is the difference between a method and a function?

A method is a function defined on a class and always takes a first parameter which is the specific instance from which the method is called.

CHAPTER 14

Using a Command Line and an Editor

In the first part of this book you used Jupyter notebooks as an interface to Python. This has a number of advantages, the strongest of which is the ability to include both code and prose in the same document. From this part of the book onwards you will explore another approach to using Python which is to use a code editor and a command line tool as a direct interface to your operating system.

In this chapter you will cover:

- Using the command line.

- Using an editor.

14.1 TUTORIAL

You will here consider a problem you have already solved in Chapter 3 but use a different interface to do so than Jupyter. The code itself will be the same. The way you run it will differ.

Rationalise the denominator of $\frac{1}{\sqrt{2}+1}$

Open a command line tool:

1. On **Windows** search for `Anaconda Prompt` (it should be available to you after installing Anaconda). See Chapter 2.

2. On **OS X** search for `terminal`. See Chapter 2.

Whether or not you are using the Windows or MacOS operating system changes the commands you need to type. First, list the directory you are currently in:

On Windows:

Command line input

```
1   $ dir
```

On MacOS:

DOI: 10.1201/9781003451860-14

Command line input

```
1  $ ls
```

This is similar to using your file explorer to view the contents in a given directory. Similarly to the way you double click on a directory in the file explorer you can navigate to a directory in the command line.

> Throughout this book, when there are commands to be typed in a command line tool they will be prefixed with a $. Do not type the $.

To do this you use the same command on both operating systems cd.

You will do this to navigate to your pfm directoy. For example if, as in Chapter 3, the pfm directory was put on the Desktop directory, you would run the following:

Command line input

```
1  $ cd Desktop
2  $ cd pfm
```

> The two statements are written under each other to denote that they are run one after the other.

You will now create a new directory:

Command line input

```
1  $ mkdir scripts
```

Inside this directory you will run the same command as before to see the contents:
On Windows:

Command line input

```
1  $ dir
```

On MacOS:

Command line input

```
1  $ ls
```

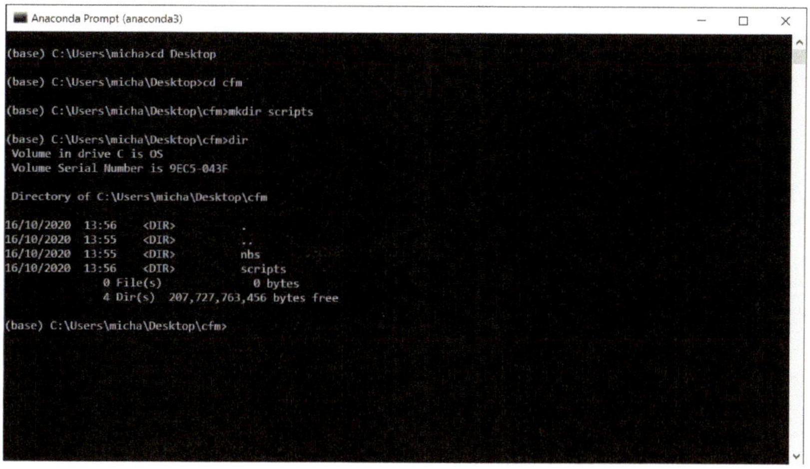

Figure 14.1 The output of `dir` on Windows.

If you have followed the steps described in Chapter 2, you will see something similar to Figure 14.1 or Figure 14.2.

Before continuing with this directory you are going to install a powerful code editor.

1. Navigate to `https://code.visualstudio.com`.

2. Download the installer making sure it is the correct one for your operating system (Windows, MacOS or Linux).

3. Run the installer.

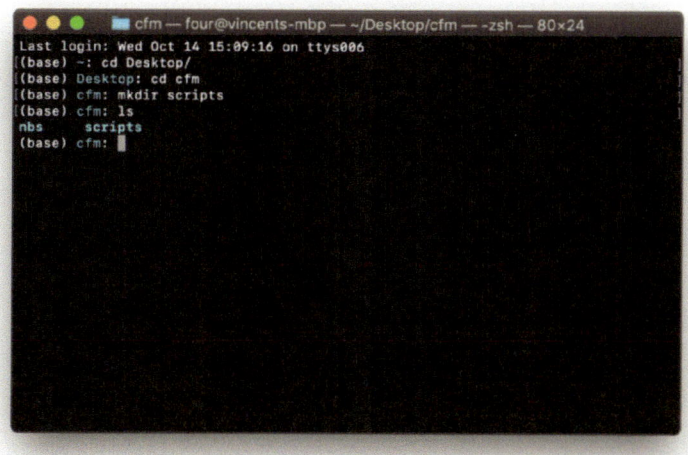

Figure 14.2 The output of `ls` on MacOS.

This code editor will offer you a different way to write Python code.

Open VS code and create a new file.

In it write the following (which corresponds to the solution of our problem):

Python file

```python
"""
This script displays the solution to the problem considered.
"""
import sympy as sym

print("Question 1:")
expression = 1 / (sym.sqrt(2) + 1)
print(sym.simplify(expression))
```

This is shown in Figure 14.3.

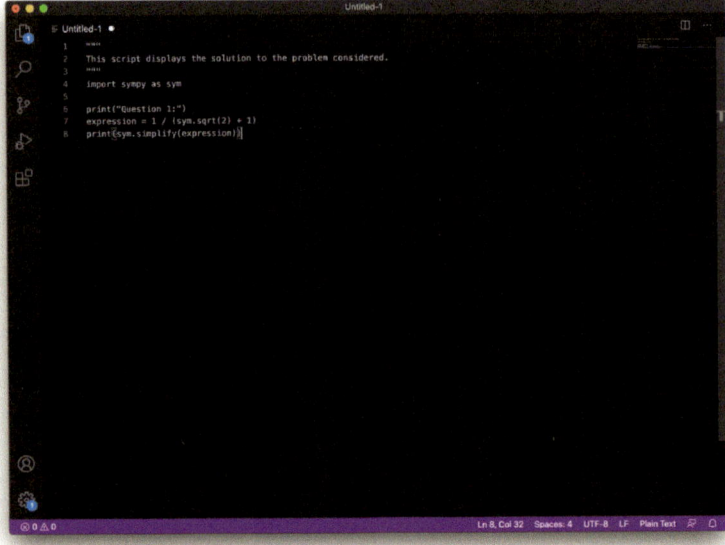

Figure 14.3   The code in VS code.

Now save this as `algebra.py` inside the `scripts` directory created earlier as shown in Figure 14.4.

VScode now recognises the Python language and adds syntax colouring. It also suggests a plugin specific for the Python language as shown in Figure 14.5. There is more information about plugins in Section 14.2.6.

All you have done so far is write the code. You now need to tell Python to run it. To do this you will use the command line and run the same command on both operating systems:

Command line input

```
$ cd scripts
```

Now confirm that the `algebra.py` file is in that directory:

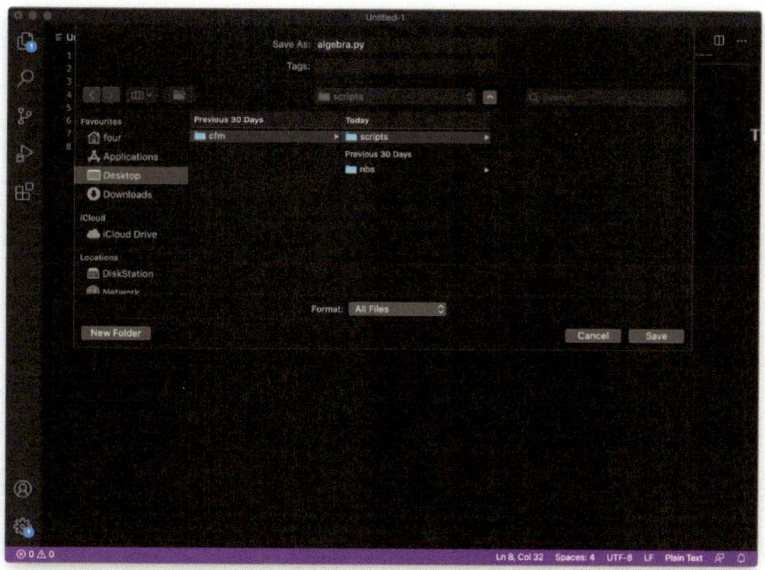

Figure 14.4   Saving file in VScode.

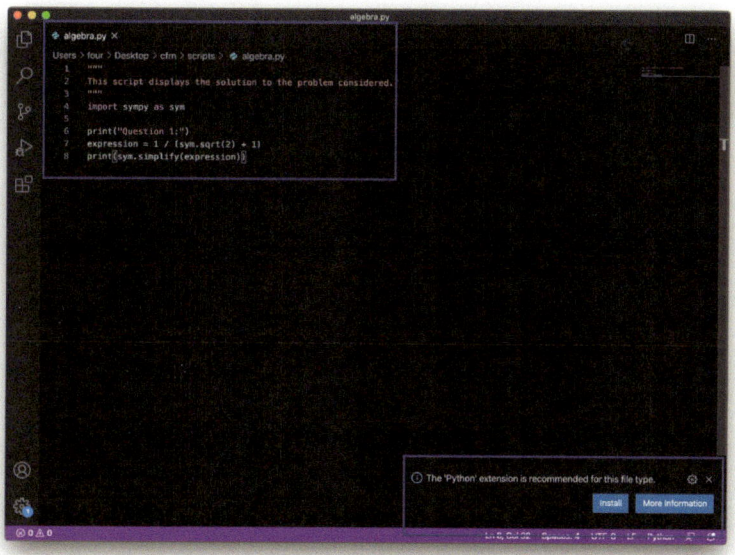

Figure 14.5   Syntax colouring and plugin suggestion.

On Windows:

> **Command line input**
>
> 1 `$ dir`

On MacOS:

> **Command line input**
>
> 1 `$ ls`

Now run the python code in that script:

> **Command line input**
>
> 1 `$ python algebra.py`

The output of this is shown in Figure 14.6

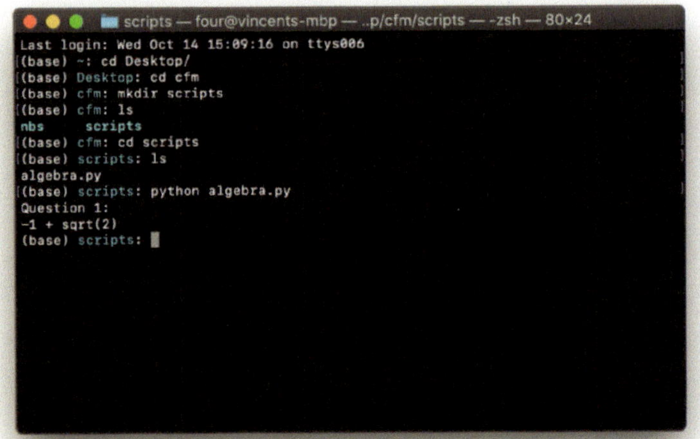

Figure 14.6   Output of running script.

14.2   HOW TO

14.2.1   Navigate directories using the command line

In the command line the `cd` command (short for "change directory") can be used to enter a given directory.

> **Usage**
>
> 1 $ cd <directory>

The target directory must be contained in the directory you are currently in.

For example, to change directory into a directory called `pfm`:

> **Command line input**
>
> 1 $ cd pfm

To go back to the previous directory use `..`:

> **Command line input**
>
> 1 $ cd ..

14.2.2   Create a new directory using the command line

In the command line the `mkdir` command (short for "make `directory`") can be used to create a new directory.

> **Usage**
>
> 1 $ mkdir <directory>

For example, to create a director called `scripts`:

> **Command line input**
>
> 1 $ mkdir scripts

14.2.3   See the contents of a directory in the command line

In the command line you can see the contents of the current directory:

- On Windows using `dir`
- On OS X using `ls`

14.2.4 Run Python code in a file

To run code in a file type `python` followed by the name of the file in the command line.

> **Usage**
>
> 1   `$ python <file.py>`

For example, to run code in a file called `main.py`:

> **Command line input**
>
> 1   `$ python main.py`

14.2.5 Run Python code without using a file or Jupyter

At the command line if you type `python` without passing a filename this will create a prompt in which you can directly write Python code.

> **Command line input**
>
> 1   `$ python`

When doing this, you see a prompt appear with `>>>`, you can directly type python code in there and press enter:

```
>>> 2 + 2
4
```

> This interface to Python is called a Read-Eval-Print-Loop and is often referred to as a REPL.

> Using `python` is the simplest of python REPLs, there are others (for example `ipython`).

This interface to Python is quite limited and should only be used for quick access to Python as a way to run simple commands.

14.2.6 Install VScode plugins

VScode is a powerful editor with a number of plugins for different languages and functionalities.

To install a particular plugin in the menu bar, click on `Code > Preferences > Extensions`.

From there you can search for a specific plugin and install it by clicking on the install button.

14.3   EXERCISES

1. Use the REPL (read-eval-print-loop) to carry out the following calculations:

   (a) $3 + 8$

   (b) $3/7$

   (c) $456/21$

   (d) $\frac{4^3+2}{2\times5} - 5^{\frac{1}{2}}$

2. Install the Python plugin for VScode.

3. Use the command line and a python script written in VScode to solve the following problems:

   (a) Find the solutions to the following equation: $x^2 - 3x + 2 = 1$.

   (b) Differentiate the following function: $f(x) = \cos(x)/4$

   (c) Find the determinant of $A = \begin{pmatrix} 1/5 & 1 \\ 1 & 1 \end{pmatrix}$.

   (d) Count the number of ways of picking 2 letters from "ABCD" where order does not matter.

   (e) Simulate the probability of picking a red token from a bag with 3 red tokens, 5 blue tokens and a yellow token.

   (f) Obtain the first five terms of the sequence defined by:

   $$\begin{cases} a_0 = 0, \\ a_1 = 2, \\ a_n = 3a_{n-1} + a_{n-2}, n \geq 2 \end{cases}$$

4. Install the `Markdown all in one` plugin for markdown in VScode and then:

   (a) Create a new file `main.md`.

   (b) Write some basic markdown in it.

   (c) Use the plugin to preview the rendered markdown.

14.4   FURTHER INFORMATION

14.4.1   Why do you need to use the `print` function with an editor?

When using a Jupyter notebook, the last line of a cell corresponds to the output of the cell and is automatically displayed. When running code written in an editor directly through the Python interpreter there is nowhere for code to be output to. Thus, you need to specifically tell it to display the code which is what the `print` statement does.

14.4.2   Can you use a Python plugin to run my code from inside my editor?

When using the Python plugin, buttons become available that let you run code without using the command line. Before using those buttons it is good to become comfortable using a command line tool to fully understand what the underlying process is. Furthermore, at times when debugging sometimes the user interface might be at fault.

14.4.3 Can I open a Jupyter notebook inside VScode?

When using the Python plugin it is actually possible to use Jupyter notebooks from within VScode. The notebooks will not look exactly the same but have the same functionality as shown in Figure 14.7.

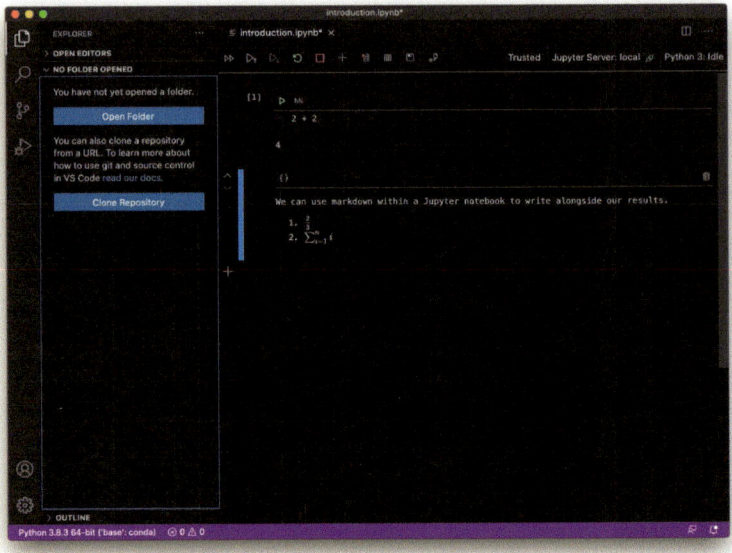

Figure 14.7   A notebook in VScode.

14.4.4 What is the difference between an Integrated Development Environment and an editor?

An **I**ntegrated **D**evelopment **E**nvironment or IDE is another type of tool used to write code. A popular one for Python is PyCharm. Generally IDEs are powerful tools designed for one specific language whereas editors are supposedly lightweight and designed to be flexible to be used with many languages.

Experiment with IDEs and/or editors to find what you prefer but throughout this book VSCode.

14.4.5 Can I use \(and \) instead of $ for LATEX?

When using Jupyter notebooks (see Section 2.5.7) or the markdown preview feature in VScode, the single $ and $$ must be used as delimiters for mathematics.

CHAPTER 15

Modularisation

This is the first of three chapters that aim to move from writing code that works to writing software. In this particular chapter you will consider how to write your code in a structured way.

In this chapter you will cover:

- Importing code from python files.

- Fragmenting code into modular components.

15.1 TUTORIAL

You will here consider a specific problem of a general type. You will not concentrate too much on the writing of the code itself. Instead this chapter concentrates on how you can write the code as software that will do more than just solve the specific problem. It will be able to be used for further problems of the same type.

Consider a chain model of the Board Game "Snakes and Ladders":

1. What is the shortest number of turns that are possible to win?

2. What is the average number of turns?

To solve this problem you will make use of the Python library numpy to carry out efficient numerical calculations.

The problem you are considering is in fact an application of a mathematical object from probability called a Chain which not be covered here; however the relevant ideas are that the probability of being in the 100th square after k turns can be written down as:

$$(\pi P^k)_{100}$$

where

$$\pi = (\underbrace{1, 0, \ldots, 0}_{100})$$

and $P \in \mathbb{R}^{100 \times 100}$; P_{ij} represents the probability of being in the ith square and going to the $j_t h$ square after rolling the dice.

There are snakes and ladders between the squares as given in Table 15.1.

The matrix P will look like:

DOI: 10.1201/9781003451860-15

TABLE 15.1 Snake and Ladders

Snake or Ladder	From	To
Ladder	3	19
Ladder	15	37
Ladder	22	42
Ladder	25	64
Ladder	41	73
Ladder	53	74
Ladder	63	86
Ladder	76	91
Ladder	84	98
Snake	11	7
Snake	18	13
Snake	28	12
Snake	36	34
Snake	77	16
Snake	47	26
Snake	83	39
Snake	92	75
Snake	99	70

$$P = \begin{pmatrix} 0 & 1/6 & 1/6 & 0 & 1/6 & 1/6 & 1/6 & 0 & 0 & \cdots & 0 \\ 0 & 0 & 1/6 & 0 & 1/6 & 1/6 & 1/6 & 1/6 & 0 & \cdots & 0 \\ \vdots & 0 & 0 & 0 & \ddots & \ddots & \ddots & \ddots & \ddots & \ddots & \vdots \\ 0 & 0 & 0 & 0 & 0 & 0 & 0 & \cdots & 0 & 0 & 1 \end{pmatrix}$$

Note that because of the ladder on square 3: $P_{14} = 0$ and $P_{1,20} = 1/6$. The first row/column of P corresponds to the state of not being on the board.

A csv file containing this matrix P can be found at https://zenodo.org/record/4236275.

To be able to answer the first question you will write a function to compute πP^k for arbitrary π, k and P:

Python file

```
def get_long_run_state(pi, k, P):
    """
    For a Markov chain with transition matrix P and starting state
    ↪ vector pi,
    obtain the state distribution after k steps.

    This is done by computing pi P ^ k

    Parameters
    ----------
    pi : array
        Starting state vector.
    k : int
        Number of iterations.
```

```
14        P : array
15            Transition matrix
16
17        Returns
18        -------
19        array
20            The state vector after k iterations
21        """
22        return pi @ np.linalg.matrix_power(P, k)
```

For the second question you are going make use of a theoretic result which is that if P is of the form:

$$P = \begin{pmatrix} Q & R \\ 0 & I \end{pmatrix}$$

In this case the **fundamental matrix** is defined by:

$$N = (I - Q)^{-1}$$

The fundamental matrix of an absorbing chain has a number of potential applications. One of which is to calculate the expected number of steps for the chain to be absorbed given by:

$$t = N\mathbb{1}$$

where $\mathbb{1}$ is a column of 1s.

To be able to code this you write a function to compute t but this requires "extracting" Q from P:

Python file

```
1   def compute_t(P):
2       """
3       For an absorbing Markov chain with transition rate matrix this
        ↪   computes the
4       vector t which gives the expected number of steps until absorption.
5
6       Note that this does not assume P is in the required format.
7
8       Parameters
9       ----------
10      P : array
11          Transition matrix
12
```

```
13      Returns
14      -------
15      array
16          Number of steps until absorption
17      """
18      indices_without_1_in_diagonal = np.where(P.diagonal() != 1)[0]
19      Q = P[indices_without_1_in_diagonal.reshape(-1, 1),
        ↪   indices_without_1_in_diagonal]
20
21      number_of_rows, _ = Q.shape
22      N = np.linalg.inv(np.eye(number_of_rows) - Q)
23      return N @ np.ones(number_of_rows)
```

You are in fact going to modularise that function. It does three things:

- Extracts the matrix Q from P;

- Computes N;

- Computes t.

All of those tasks could be useful in their own right so you are going to break up that function into three separate functions:

Python file

```
1   def extract_Q(P):
2       """
3       For an absorbing Markov chain with transition rate matrix P this
        ↪   computes the
4       matrix Q.
5
6       Note that this does not assume that P is in the required format. It
7       identifies the rows and columns that have a 1 in the diagonal and
        ↪   removes
8       them.
9
10      Parameters
11      ----------
12      P : array
13          Transition matrix
14
15      Returns
16      -------
17      array
18          The matrix Q
19      """
20      indices_without_1_in_diagonal = np.where(P.diagonal() != 1)[0]
```

```
21      Q = P[indices_without_1_in_diagonal.reshape(-1, 1),
        ↪  indices_without_1_in_diagonal]
22      return Q
23
24
25  def compute_N(Q):
26      """
27      For an absorbing Markov chain with transition rate matrix P that
        ↪  gives
28      matrix Q this computes the fundamental matrix N.
29
30      Parameters
31      ----------
32      Q : array
33          The matrix Q obtained from P
34
35      Returns
36      -------
37      array
38          The funamental matrix N
39      """
40      number_of_rows, _ = Q.shape
41      N = np.linalg.inv(np.eye(number_of_rows) - Q)
42      return N
```

This now allows you to redefine `compute_t` in a simpler way:

Python file

```
1   def compute_t(P):
2       """
3       For an absorbing Markov chain with transition rate matrix this
        ↪  computes the
4       vector t which gives the expected number of steps until absorption.
5
6       Note that this does not assume P is in the required format.
7       """
8       Q = extract_Q(P)
9       N = compute_N(Q)
10      number_of_rows, _ = Q.shape
11      return N @ np.ones(number_of_rows)
```

All the code you have written so far is generic in nature so would be better placed somewhere that it could be used for any other project.

You are going to put these three functions (and the necessary `import numpy as np` statement) in an `absorption.py` file as can be seen in Figure 15.1.

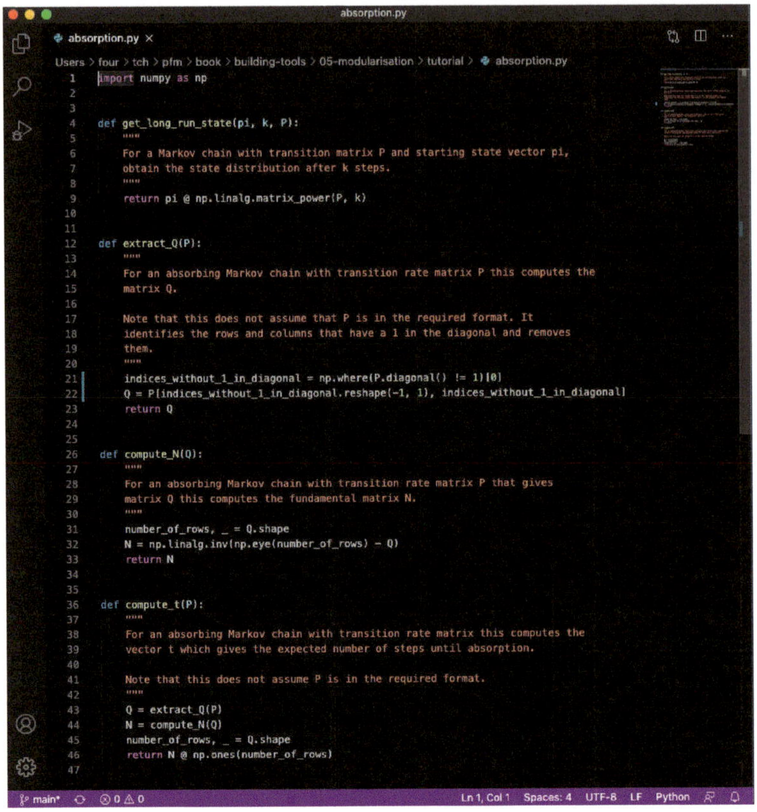

Figure 15.1   The three modularised function in a python file.

You will now use everything you have done so far:

- Download, and extract the data available at `https://zenodo.org/record/4236275`. Put the `main.csv` file in the `snakes_and_ladders` directory.

- Open a Jupyter notebook in the `snakes_and_ladders` directory and call it `main.ipynb`.

This should look like the following:

```
snakes and ladders/
├── absorption.py
├── main.csv
├── main.ipynb
```

You can now use all of the code you have written in the notebook, first you can import the functions in `absorption.py` like any other python library:

Jupyter input

```
1   import absorption
```

You will also import `numpy` and use it to read the data file:

Jupyter input

```
1  import numpy as np
2
3  P = np.loadtxt("main.csv", delimiter=",")
```

The above commands work because the 3 files are all in the same directory.

Now to compute the shortest number of turns:

Jupyter input

```
1  k = 1
2  pi = np.zeros(101)
3  pi[0] = 1
4  while absorption.get_long_run_state(pi, k, P)[-1] == 0:
5      k += 1
6  k
```

It is possible to arrive at the last square in six turns.
Now to compute the average length of the game:

Jupyter input

```
1  t = absorption.compute_t(P)
2  t[0]
```

43.49196169497175

15.2   HOW TO

15.2.1   Import code from Python files

Given a `<file.py>` file in a directory any other python process in the same directory can import that file as it would a normal library.

Usage

```
1  import <file>
```

At this stage it is possible to uses any python object (a `function`, a `class`, a `variable`) by referring to the `<file.py>` as a library:

> ### Usage
>
> ```
> 1  <file>.function
> 2  <file>.class
> 3  <file>.variable
> ```

15.2.2 Break up code into modular components

The aim of Modularising code is to identify specific components of the code that can be isolated from the rest. In practice this means writing multiple functions that use the correct inputs and outputs in sequence for an overall goal.

Often this allows you to write a more comprehensive docstring that explains specific parts of the implemented process. As an example, consider the problem of wanting to pay a shared bill after applying a tip, the following function will do this:

> ### Jupyter input
>
> ```python
> 1  def add_tip_and_get_bill_share(total, tip_proportion,
>    ↪    number_of_payers):
> 2      """
> 3      This returns the share of a bill to be paid by `number_of_payers`
> 4      ensuring the total paid includes a tip.
> 5
> 6      Parameters
> 7      ----------
> 8      total : float
> 9          The total amount of the bill
> 10     tip_proportion : float
> 11         The proportion of the bill that should be added as a tip (a
>    ↪        number
> 12         between 0 and 1)
> 13     number_of_payers : int
> 14         The number of people sharing the bill
> 15
> 16     Returns
> 17     -------
> 18     float
> 19         The amount each person should contribute
> 20     """
> 21     tip_amount = tip_proportion * total
> 22     total += tip_amount
> 23     return total / number_of_payers
> ```

You can check that this works:

```
1  add_tip_and_get_bill_share(total=100, tip_proportion=0.2,
   ↪  number_of_payers=6)
```

```
20.0
```

An improvement of the above would be:

```
1  def add_tip(total, tip_proportion):
2      """
3      This adds the given proportion to a bill total.
4
5      Note that tip_proportion is a number between 0 and 1. A
       ↪  tip_proportion of 0
6      corresponds to no tip and a tip_proportion of 1 corresponds to
       ↪  paying the
7      total twice.
8
9      Parameters
10     ----------
11     total : float
12         The total amount of the bill
13     tip_proportion : float
14         The proportion of the bill that should be added as a tip (a
           ↪  number
15         between 0 and 1)
16
17     Returns
18     -------
19     float
20         The total value of the bill (including tip)
21     """
22     tip_amount = tip_proportion * total
23     return total + tip_amount
24
25
26  def get_bill_share(total, number_of_payers):
27      """
28      This returns the share of a bill by dividing the total by the
        ↪  number of
29      payers.
30
31      Parameters
32      ----------
```

```
33        total : float
34            The total amount of the bill
35        number_of_payers : int
36            The number of people sharing the bill
37
38        Returns
39        -------
40        float
41            The amount each person should contribute
42        """
43        return total / number_of_payers
```

Then to use the above you would be able to explicitly write out each step which ensures that there is clarity in what is being done:

Jupyter input

```
1   total = add_tip(total=100, tip_proportion=0.2)
2   get_bill_share(total=total, number_of_payers=6)
```

```
20.0
```

15.3 EXERCISES

1. Use the code written in Section 15.1 to obtain the average time until absorption from the first state of the chain with the following transition matrices:

    (a) $P = \begin{pmatrix} 1/2 & 1/2 \\ 0 & 1 \end{pmatrix}$

    (b) $P = \begin{pmatrix} 1/2 & 1/4 & 1/4 \\ 1/3 & 1/3 & 1/3 \\ 0 & 0 & 1 \end{pmatrix}$

    (c) $P = \begin{pmatrix} 1 & 0 \\ 1/2 & 1/2 \end{pmatrix}$

    (d) $P = \begin{pmatrix} 1/2 & 1/4 & 1/4 \\ 1/3 & 1/3 & 1/3 \\ 1/5 & 0 & 4/5 \end{pmatrix}$

2. Modularise the following code by creating a function flip_coin that takes a probability_of_selecting_heads variable.

```
     Jupyter input

1    import random
2
3    def sample_experiment(bag):
4        """
5        This samples a token from a given bag and then
6        selects a coin with a given probability.
7
8        If the sampled token is red then the probability
9        of selecting heads is 4/5 otherwise it is 2/5.
10
11       This function returns both the selected token
12       and the coin face.
13       """
14       selected_token = random.choice(bag)
15
16       if selected_token == "Red":
17           probability_of_selecting_heads = 4 / 5
18       else:
19           probability_of_selecting_heads = 2 / 5
20
21       if random.random() < probability_of_selecting_heads:
22           coin = "Heads"
23       else:
24           coin = "Tails"
25       return selected_token, coin
```

3. Modularise the following code by writing two further functions:

   - get_potential_divisors: A function to return the integers between 2 and $\sqrt{N}$ for a given N.

   - is_divisor: A function to check if $n|N$ ("n divides N") for given n, N.

```
     Jupyter input

1    import math
2
3    def is_prime(N):
4        """
5        Checks if a number N is prime by checking all that positive
         ↪  integers
6        numbers less sqrt(N) than it that divide it.
7
8        Note that if N is not a positive integer great than 1 then it
         ↪  does not
9        check: it returns False.
```

```
10      """
11      if N <= 1 or type(N) is not int:
12          return False
13      for potential_divisor in range(2, int(math.sqrt(N)) + 1):
14          if (N % potential_divisor) == 0:
15              return False
16      return True
```

Confirm your work by comparing to the results of using `sympy.isprime`.

4. Write a `stats.py` file with two functions to implement the calculations of mean and population variance.

Note that the mean is defined by:

$$\bar{x}\frac{\sum_{i=1}^{N} x_i}{N}$$

The population variance is defined by:

$$\sigma^2 = \frac{\sum_{i=1}^{N}(x_i - \bar{x})^2}{N}$$

Use your functions to compute the mean and population variance of the following collections of numbers:

(a) $S_1 = (1, 2, 3, 4, 4, 4, 4, 4)$

(b) $S_1 = (1)$

(c) $S_1 = (1, 1, 1, 2, 3, 4, 4, 4, 4, 4)$

Compare your results to the results of using the `statistics.mean` and `statistics.pvariance`.

15.4   FURTHER INFORMATION

15.4.1   Why modularise?

Best practice when writing code is breaking up of code into modular parts. Here is a guiding principle described in [3]:

> "Code should be obvious. When someone needs to make a change, they should be able to find the code to be changed easily and make the change quickly without introducing any errors."

While this guiding principle is ambiguous and all concepts related to clean code writing and refactoring are not things that can be covered in this book, one specific principle is the one referred to in [4]:

> "Functions should do one thing. They should do it well. They should do it only."

In some texts on code architecture you will see arbitrary rules about how many lines of code should be in a given function. Having a function with 10 or more lines of code might indicate that it can be modularised. **However**, it is not recommended to follow such rules strictly. Sometimes they might add more complexity than they remove. Make your code clear and ensure your functions do one thing well and one thing only.

15.4.2 Why do I get an import error?

The most probable explanation for this is that you are importing a file that is not in the same directory or that you have not imported the file with the correct name.

As an example, if your code is in a `library` directory but that your notebook is in a **different** directory then you will get an error as shown below:

Jupyter input

```
1  import library
```

```
ModuleNotFoundError                     Traceback (most recent call last)
Cell In[1], line 1
----> 1 import library

ModuleNotFoundError: No module named 'library'
```

Similarly if you perhaps incorrectly saved your `library.py` file with a typo in the name such as: `librery.py` then you would get the same error.

15.4.3 How do I make my file importable from other directories?

This falls under the subject matter of "packaging". This is not covered in this book.

CHAPTER 16

Documentation

This is the second of three chapters that aims to move from writing code that works to writing software. In this particular chapter you will consider how to write documentation for your code.

In this chapter you will cover:

- Using the Diátaxis framework for documentation [9].

16.1 TUTORIAL

In this tutorial you will write documentation for the code you wrote in Section 15.1.

You start by creating a new file in VScode called README.md.

You will be writing your documentation in markdown.

Start by writing the title of your library and a quick single sentence description.

Markdown input

```
1   # Absorption
2
3   Functionality to study the absorbing Markov chains.
```

16.1.1 Writing a tutorial

You will then write your first section which is a **tutorial**.

The goal of a tutorial is to provide a hands on introduction and demonstration of the software.

DOI: 10.1201/9781003451860-16

Markdown input

```
In this tutorial we will see how to use `absorption` to study an
  absorbing
Markov chain. For some background information on absorbing Markov
  chains we
recommend: <https://en.wikipedia.org/wiki/Absorbing_Markov_chain>.

Given a transition matrix $P$ defined by:

$$
p = \begin{pmatrix}
    1/2 & 1/4 & 1/4\\
    1/3 & 1/3 & 1/3\\
    0   & 0   & 1
    \end{pmatrix}
$$

We will start by seeing how the chain evolves over time by starting
  with an
initial vector $\pi=(1,0,0)$. In the next code snippet we will import
  the
necessary libraries and create both $P$ and $\pi$:

```python
import numpy as np

import absorption

pi = np.array([1, 0, 0])
P = np.array([[1 / 2, 1 / 4, 1 / 4], [1 / 3, 1 / 3, 1 / 3], [0, 0, 1]])
```

We now see how the vector π changes over time:

```python
for k in range(10):
    print(absorption.get_long_run_state(pi, k, P))
```

This will give:

```
[1. 0. 0.]
[0.5  0.25 0.25]
[0.33333333 0.20833333 0.45833333]
[0.23611111 0.15277778 0.61111111]
```
```

```
43    [0.16898148 0.1099537  0.72106481]
44    [0.12114198 0.0788966  0.79996142]
45    [0.08686986 0.05658436 0.85654578]
46    [0.06229638 0.04057892 0.8971247 ]
47    [0.0446745 0.0291004 0.9262251]
48    [0.03203738 0.02086876 0.94709386]
49    ```

50

51    We see that, as expected over time the probability of being in the
      ↪  third state,
52    which is absorbing, increases.

53

54    We can also use `absorption` to get the average number of steps until
55    absorption from each state:

56

57    ```python
58    absorption.compute_t(P)
59    ```

60

61    This gives:

62

63    ```
64    array([3.66666667, 3.33333333])
65    ```

66

67    We see that the expected amounts of steps from the first state is
      ↪  slightly more
68    than from the second.
```

This **tutorial** section allows newcomers to your code to see how it is intended to be used.

16.1.2 Writing the how-to guides

In the next section you will write a series of **how to** guides, this is targeted at someone who has perhaps worked through the tutorial already and wants to directly know how to do a specific task.

Directly underneath what you have written so far write:

Markdown input

```
1
2    ## How to guides
3
4    ### How to compute the long run state of a system after a given number
     ↪  of steps
5
```

```
 6  Given a transition matrix $P$ and a state vector $\pi$, the state of
 ↪    the system
 7  after $k$ steps is given by:
 8
 9  ```python
10  import numpy as np
11
12  import absorption
13
14  pi = np.array([0, 1, 0])
15  P = np.array([[1 / 3, 1 / 3, 1 / 3], [1 / 4, 1 / 4, 1 / 2], [0, 0, 1]])
16  absorption.get_long_run_state(pi=pi, k=10, P=P)
17  ```
18
19  This gives:
20
21  ```
22  array([0.0019552, 0.0019552, 0.9960896])
23  ```
24
25  ### How to extract the transitive state transition sub matrix $Q$
26
27  Given a transition matrix $P$, the sub matrix $Q$ that
28  corresponds to the transitions between transitive (i.e. not absorbing)
 ↪    states can
29  be extracted:
30
31  ```python
32  import numpy as np
33
34  import absorption
35
36  P = np.array([[1 / 3, 1 / 3, 1 / 3], [0, 1, 0], [1 / 4, 1 / 4, 1 / 2]])
37  absorption.extract_Q(P=P)
38  ```
39
40  This gives:
41
42  ```
43  array([[0.33333333, 0.33333333],
44         [0.25      , 0.5       ]])
45  ```
46
47  ### How to compute the fundamental matrix $N$
48
49  Given a transition matrix $P$, the fundamental matrix $N$
50  can be obtained:
51
```

```python
52  ```python
53  import numpy as np
54
55  import absorption
56
57  P = np.array([[1 / 3, 1 / 3, 1 / 3], [0, 1, 0], [1 / 4, 1 / 4, 1 / 2]])
58  Q = absorption.extract_Q(P=P)
59  absorption.compute_N(Q=Q)
60  ```
61
62  This gives:
63
64  ```
65  array([[2.        , 1.33333333],
66         [1.        , 2.66666667]])
67  ```
68
69  ### How to compute the average steps until absorption
70
71  Given a transition matrix $P$ and a state vector $\pi$, the average
    ↪ number of
72  steps until absorption from all states can be obtained:
73
74  ```python
75  import numpy as np
76
77  import absorption
78
79  P = np.array([[1 / 3, 1 / 3, 1 / 3], [0, 1, 0], [1 / 4, 1 / 4, 1 / 2]])
80  absorption.compute_t(P=P)
81  ```
82
83  This gives:
84
85  ```
86  array([3.33333333, 3.66666667])
87  ```
```

This **how to** section is an efficient collection of instructions to be able to carry out specific tasks made possible by the software.

16.1.3 Writing the explanations section

In the next section you will write the **explanations** which aims to give more in depth understanding not necessarily directly related to the code.

Markdown input

```
1   ## Explanation
2
3   ### Brief overview of absorbing Markov chains
4
5   A Markov chain with a given transition matrix $P$ is a system that
    ↪  moves from
6   state to state randomly with the probabilities given by $P$.
7
8   For example:
9
10  $$
11  P = \begin{pmatrix}
12        1 / 3 & 1 / 3 & 1 / 3 \\
13        0     & 1     & 0     \\
14        1 / 4 & 1 / 4 & 1 / 2
15     \end{pmatrix}
16  $$
17
18  The corresponding Markov chain has 3 states and:
19
20  - $P_{11}=1/3$ means that when in state 1 there is a 1/3 chance of
    ↪  staying in
21    state 1.
22  - $P_{23}=0$ means that when in state 2 there is a 0 chance of staying
    ↪  in
23    state 1.
24  - $P_{22}=$ actually implies that once we are in state 2 that the only
    ↪  next
25    state is state 2.
26
27  In general:
28
29  $$
30      P_{ij} > 0 \text{ for all }ij
31  $$
32
33  $$
34      \sum_{j=0}^{|P|} P_{ij} = 1 \text{ for all }i
35  $$
36
37  If $P_{ii}=1$ then state $i$ is an absorbing state from which no
    ↪  further changes
38  can occur.
39
40  In the case of absorbing markov chains there are a number of
    ↪  quantities that can
```

41 be measured.

42

43 ### Calculating the state after a given number of iterations

44

45 Given a vector that describes the state of the system π and a
    ↪ transition
46 matrix P, the state of the system after k iterations will be given
    ↪ by the
47 following vector:

48

49 $$
50   \pi P ^ k
51 $$

52

53 ### The canonical form of the transition matrix

54

55 A transition matrix P is written in its canonical form when it can
    ↪ be written
56 as:

57

58 $$
59   P =
60   \left(\begin{array}{c|c}
61   Q & R \\\hline
62   0 & I
63   \end{array}\right)
64 $$

65

66 Where Q is the matrix of transitions between non absorbing states.

67

68 For example, the canonical form of P would be:

69

70 $$
71   \begin{pmatrix}
72   1 / 3 & 1 / 3 & 1 / 3 \\
73   1 / 4 & 1 / 2 & 1 / 4 \\
74   0   & 0   & 1   \\
75   \end{pmatrix}
76 $$

77

78 which would give:

79

80 $$
81 Q = \begin{pmatrix}
82   1 / 3 & 1 / 3 \\
83   1 / 4 & 1 / 2
84   \end{pmatrix}
85 $$

```
86
87   ### The fundamental matrix
88
89   Given $Q$ the fundamental matrix $N$ is defined as:
90
91   $$N = (I - Q) ^{-1}$$
92
93   $N_{ij}$ corresponds to the expected number of times the chain will be
     ↪  in state
94   $j$ given that it started in state $i$.
95
96   ### The expected number of steps until absorption.
97
98   Given $N$, the expected number of steps until absorption is given by
     ↪   the vector:
99
100  $$
101  t = N \mathbb{1}
102  $$
103
104  where $\mathbb{1}$ denotes a vector of 1s.
```

This **explanations** section gives background reading as to how the code works.

16.1.4   Writing the reference section

In the next section you will write the **reference** which aims to be a concise collection of reference material.

Markdown input

```
1    ## Reference
2
3    ### List of functionality
4
5    The following functions are written in `absorption`:
6
7    - `get_long_run_state`
8    - `extract_Q`
9    - `compute_N`
10   - `compute_t`
11
12   ### Bibliography
13
14   The wikipedia page on absorbing Markov chains gives a good overview of
     ↪   the
15   topic: <https://en.wikipedia.org/wiki/Absorbing_Markov_chain>
```

```
16
17   The following text is a recommended reference on Markov chains:
18
19   > Stewart, William J. Probability, Markov chains, queues, and
     ↪ simulation: the
20   > mathematical basis of performance modelling. Princeton university
     ↪ press, 2009.
```

Figure 16.1 shows the start of the markdown file in VScode alongside the preview. Note that the `Markdown all in one` plugin ensures that the mathematics is rendered, see Section 14.2.6 for information on installing plugins.

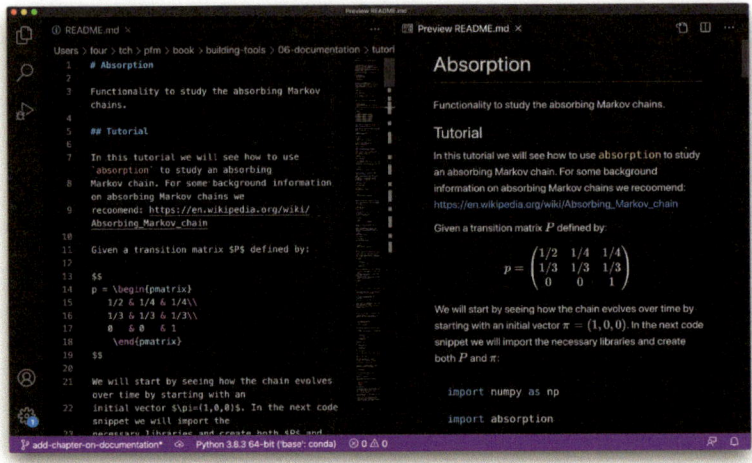

Figure 16.1 The `README.md` file in VScode with the rendered preview (using the `Markdown all in one` plugin).

16.2 HOW TO

16.2.1 Write documentation

Follow the Diátaxis framework [9] for documentation.

This involves separating your documentation into four different sections based on separate aims for readers.

- Tutorial: for learning.

- How to guides: to achieve a specific goal.

- Explanation: to understand.

- Reference: to find information.

16.2.1.1 Write a tutorial

A tutorial should include step by step instructions with expected behaviours. This should not focus on any deeper explanation.

An analogy of this is teaching a young toddler to build a toy train track. They do not need to know the physics related to how the train will go through the track. They need only to see how to lay the track pieces.

16.2.1.2 How to write a how to guide

A how to guide should provide a quick and to the point description of how to solve a specific problem.

An analogy of this would be a recipe. The recipe will not necessarily explain how to chop an onion and/or why you are chopping an onion. It will tell you how to chop an onion as a step of cooking a particular meal.

16.2.1.3 How to write an explanation section

The explanation section should provide a deeper understanding of the concepts under the code.

An analogy of this again related to a recipe would be a book on the chemistry of taste and why a chopped onion adds a specific type of flavour to a meal.

16.2.1.4 How to write a reference section

The reference section should provide an overview of the specific tools, commands and indeed place for background reading as well.

16.2.2 Write markdown

16.2.2.1 How to include section headers in markdown

To include a section header in markdown use #. The number of # corresponds to the level of the section header.

Markdown input

```
1   # Section
```

For example:

Markdown input

```
1   # The absorption library
2
3   Functionality to study the absorbing Markov chains.
```

16.2.2.2 Include code in markdown

To include code in markdown use three ' marks followed by the name of the language:

> **Markdown input**
>
> ```
> 1  ```<language>
> 2
> 3  <code>
> 4  ```
> ```

For example:

> **Markdown input**
>
> ```
> 1  ```python
> 2  import sympy as sym
> 3
> 4  x = sym.Symbol(''x'')
> 5  ```
> ```

It is also possible to include code in markdown using an indented block:
For example:

> **Markdown input**
>
> ```
> 1  Here is some code:
> 2
> 3      import sympy as sym
> 4
> 5      x = sym.Symbol(''x'')
> ```

> Using an indented block does not allow you to specify the language and can lead to mistake when combining with other nested statement.

16.2.2.3 How to include a hyperlink in markdown

To include a hyperlink in markdown use []() language:

> **Markdown input**
>
> ```
> 1  [text](url)
> ```

For example:

Markdown input

```
1   The [Online Encyclopedia of Integer Sequences](https://oeis.org) is a
    ↪   good resources for studying
2   resources.
```

16.2.2.4   *How to include an image in markdown*

To include an image in markdown use :

Markdown input

```
1   ![caption](path)
```

For example:

Markdown input

```
1   Here is an image:
2
3   ![An image](image.jpg)
```

If the image file is not located in the same directory as the markdown file, the path to the file must be correct.

16.3   EXERCISES

Write documentation for the `statistics.py` file written in the exercises of Chapter 15.

16.4   FURTHER INFORMATION

16.4.1   What is documentation?

Documentation can have many different interpretations. A good definition is given in [5].

> "The process of transferring valuable knowledge to other people now and also to people in the future."

It is important to realise that the target of the documentation can be the writer of the software itself at a future date.

There are two types of documentation:

- **Internal documentation** which includes things like docstrings and a good choice of variable names.

- **External documentation** which includes things like the README.md and other separate documentation.

For a software project to be well documented, it needs **both** internal and external documentation.

In [5] there are four properties of documentation:

- Reliable: it needs to be accurate.

- Low effort: it should require minimal effort to update when changes are made to the code base.

- Collaborative: it should be a tool from which collaboration can occur.

- Insightful: it should give information not only to be able to use the code but also to understand specific reasons why certain decisions have been made as to its design.

16.4.2   What is the purpose of the four separate sections in documentation?

As discussed in [9]:

"Tutorials are lessons that take the reader by the hand through a series of steps to complete a project of some kind. They are what your project needs in order to show a beginner that they can achieve something with it."

"How-to guides take the reader through the steps required to solve a real-world problem"'

"Reference guides are technical descriptions of the machinery and how to operate it."

"Explanation, or discussions, clarify and illuminate a particular topic. They broaden the documentation's coverage of a topic."

It is natural when describing a project for the boundaries between these four topics to become fuzzy. Thus, having them explicitly in four separate sections ensures the reader is able to specifically find what they need.

16.4.3   What alternatives are there to writing documentation in the README.md file?

A single README.md file is a good way to start documenting code. However, as a project grows it could be beneficial to use some other tools. One such example of this is to use sphinx. This uses a different markup language called **restructured text** and helps build more complex documents but also interfaces to the code itself if necessary. For example, it is possible to include the code docstrings directly in the documentation (a good way of adding to the reference section).

Testing

This is the third and last chapter that shows how to move from writing code that works to writing software. In this particular chapter you will consider how to write automated tests for your software.

In this chapter you will cover:

- Assert statements to test code.

- Testing documentation.

17.1 TUTORIAL

In this tutorial you will write code to ensure the correctness of the software you have written in the tutorials of Chapters 15 and 16.

The software for `absorption.py` is in fact across two separate files:

- `absorption.py`: the source code. You will check this using **unit tests**

- `README.md`: the documentation. You will check this using **doctests**.

17.1.1 Writing tests for code

Recalling the code written in `absorption.py` in Section 15.1, there are four functions that need to be tested:

- `get_long_run_state`

- `extract_Q`

- `compute_N`

- `compute_t`

In the directory that contains `absorption.py` create a new Python file called: `test_absorption.py`.

Write the following functions to test each of the functions in `absorption.py`:

Python file

```python
1   import numpy as np
2
3   import absorption
4
5   def test_long_run_state_for_known_number_of_states():
6       """
7       This tests the `long_run_state` for a small example matrix
8       """
9       pi = np.array([1, 0, 0])
10      P = np.array([[1 / 2, 1 / 4, 1 / 4], [1 / 3, 1 / 3, 1 / 3], [0, 0,
        ↪ 1]])
11      pi_after_5_steps = absorption.get_long_run_state(pi=pi, k=5, P=P)
12      assert np.array_equal(pi_after_5_steps, pi @
        ↪ np.linalg.matrix_power(P, 5)), "Did not get expected result
        ↪ for pi after 5 steps"
13
14
15  def test_long_run_state_when_starting_in_absorbing_state():
16      """
17      This tests the `long_run_state` for a small example matrix.
18
19      In this test we start in the absorbing state, the state vector
        ↪ should not
20      change.
21      """
22      pi = np.array([0, 0, 1])
23      P = np.array([[1 / 2, 1 / 4, 1 / 4], [1 / 3, 1 / 3, 1 / 3], [0, 0,
        ↪ 1]])
24      pi_after_5_steps = absorption.get_long_run_state(pi=pi, k=5, P=P)
25      assert np.array_equal(pi_after_5_steps, pi)
26
27
28  test_long_run_state_for_known_number_of_states()
29  test_long_run_state_when_starting_in_absorbing_state()
```

The two functions you have written do not include a `return` statement but instead include an `assert` statement. An `assert` is followed by two values separated by a comma:

1. A boolean that is to be `True` or `False`.

2. A string that is output if the boolean is "False".

To run those tests, run the script at the command line:

Command line input

```
1  $ python test_absorption.py
```

When running the tests if everything has been done correctly, there will be no output: the 2 functions have been called and the assertions have **passed**. See Figure 17.1.

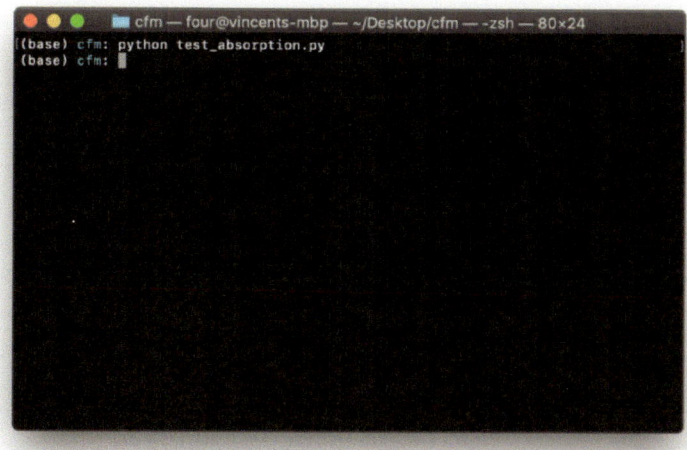

Figure 17.1   Running the tests with no errors.

For each of the 4 functions in `absorption.py`, you can now add further tests and ensure they are also called at the end. The full `test_absorption.py` file should look like:

Python file

```python
1   import numpy as np
2
3   import absorption
4
5   def test_long_run_state_for_known_number_of_states():
6       """
7       This tests the `long_run_state` for a small example matrix
8       """
9       pi = np.array([1, 0, 0])
10      P = np.array([[1 / 2, 1 / 4, 1 / 4], [1 / 3, 1 / 3, 1 / 3], [0, 0,
        ↪  1]])
11      pi_after_5_steps = absorption.get_long_run_state(pi=pi, k=5, P=P)
12      assert np.array_equal(pi_after_5_steps, pi @
        ↪  np.linalg.matrix_power(P, 5)), "Did not get expected result
        ↪  for pi after 5 steps"
```

```
13
14
15  def test_long_run_state_when_starting_in_absorbing_state():
16      """
17      This tests the `long_run_state` for a small example matrix.
18
19      In this test we start in the absorbing state, the state vector
        ↪  should not
20      change.
21      """
22      pi = np.array([0, 0, 1])
23      P = np.array([[1 / 2, 1 / 4, 1 / 4], [1 / 3, 1 / 3, 1 / 3], [0, 0,
        ↪  1]])
24      pi_after_5_steps = absorption.get_long_run_state(pi=pi, k=5, P=P)
25      assert np.array_equal(pi_after_5_steps, pi)
26
27
28  def test_extract_Q():
29      """
30      This tests that the submatrix Q can be extracted from a given
        ↪  matrix P.
31      """
32      P = np.array([[1 / 2, 1 / 4, 1 / 4], [1 / 3, 1 / 3, 1 / 3], [0, 0,
        ↪  1]])
33      Q = absorption.extract_Q(P)
34      expected_Q = np.array([[1 / 2, 1 / 4], [1 / 3, 1 / 3]])
35      assert np.array_equal(Q, expected_Q), f"The expected Q did not
        ↪  match, the code obtained {Q}"
36
37
38  def test_compute_N():
39      """
40      This tests the computation of the fundamental matrix N
41      """
42      P = np.array([[1 / 2, 1 / 4, 1 / 4], [1 / 3, 1 / 3, 1 / 3], [0, 0,
        ↪  1]])
43      Q = absorption.extract_Q(P)
44      N = absorption.compute_N(Q)
45      expected_N = np.array([[8 / 3, 1], [4 / 3, 2]])
46      assert np.allclose(N, expected_N), f"The expected N did not match,
        ↪  the code obtained {N}"
47
48
49  def test_compute_t():
50      """
51      This tests the computation of the number of steps until absorption
        ↪  t.
52      """
```

```
53    P = np.array([[1 / 2, 1 / 4, 1 / 4], [1 / 3, 1 / 3, 1 / 3], [0, 0,
      ↪   1]])
54    t = absorption.compute_t(P)
55    expected_t = np.array([11 / 3, 10 / 3])
56    assert np.allclose(t, expected_t), f"The expected t did not match,
      ↪   the code obtained {t}"
57
58
59  test_long_run_state_for_known_number_of_states()
60  test_long_run_state_when_starting_in_absorbing_state()
61  test_extract_Q()
62  test_compute_N()
63  test_compute_t()
```

The `numpy.array_equal` and `numpy.allclose` compare equality of arrays. They return `True` or `False` depending on whether the two passed arrays are equal or approximately equal (respectively).

`numpy.allclose` should be used when comparing numpy arrays that might be different due to numerical imprecision.

You can experiment by changing some of the code or the tests and see the way the tests fail. See Figure 17.2 where the following specific error has been introduced into `absorption.py`: `P.diagonal() == 1` is incorrect and should be `P.diagonal() != 1`.

As and when you add more features to `absorption.py` you will also add tests.

Software is compromised of both code and documentation. So far you have tested your code, now you will test your documentation.

17.1.2 Testing documentation

To be able to check the Python code written in the documentation (see Chapter 16) is correct, you need to write the code using a specific format:

- `>>>` to denote python code

- `...` to denote secondary lines of multi-line python code.

- Nothing to denote the expected output.

As an example, in Section 16.1, you have written:

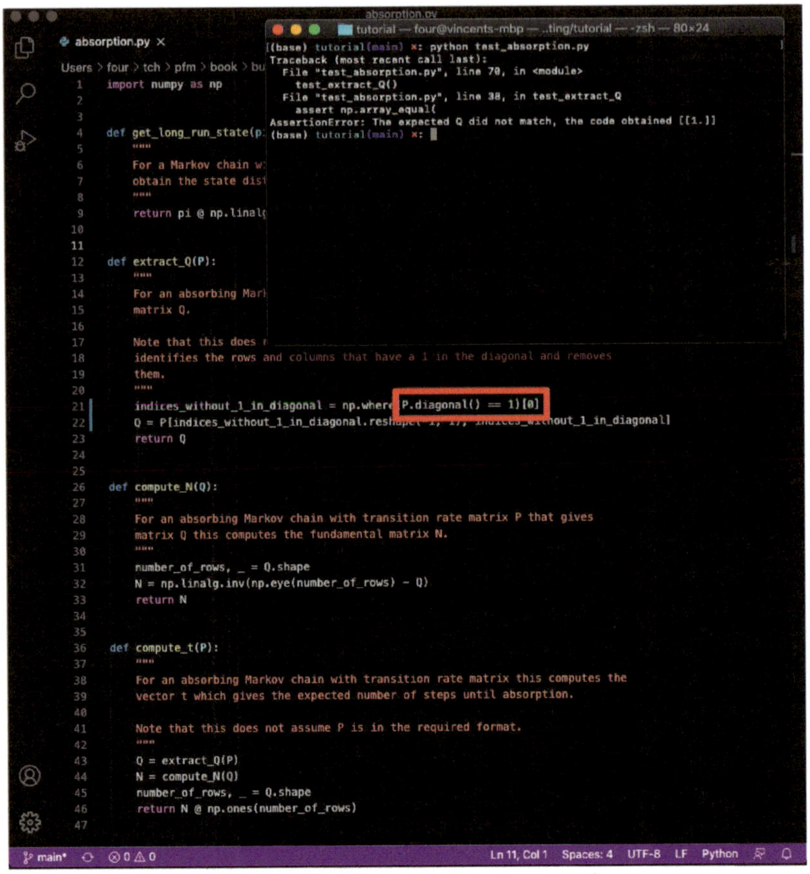

Figure 17.2 Running the tests with an error in the source code.

Markdown input

```
1   ```python
2   import numpy as np
3
4   import absorption
5
6   pi = np.array([1, 0, 0])
7   P = np.array([[1 / 2, 1 / 4, 1 / 4], [1 / 3, 1 / 3, 1 / 3], [0, 0, 1]])
8   ```
9
10  We now see how the vector $\pi$ changes over time:
11
12  ```python
13  for k in range(10):
14      print(absorption.get_long_run_state(pi, k, P))
15  ```
16
```

```
17  This will give:
18
19  ```
20  [1. 0. 0.]
21  [0.5  0.25 0.25]
22  [0.33333333 0.20833333 0.45833333]
23  [0.23611111 0.15277778 0.61111111]
24  [0.16898148 0.1099537  0.72106481]
25  [0.12114198 0.0788966  0.79996142]
26  [0.08686986 0.05658436 0.85654578]
27  [0.06229638 0.04057892 0.8971247 ]
28  [0.0446745 0.0291004 0.9262251]
29  [0.03203738 0.02086876 0.94709386]
30
31  ```
```

You will modify the above to be:

Markdown input

```python
1   ```python
2   >>> import numpy as np
3   >>> import absorption
4   >>> pi = np.array([1, 0, 0])
5   >>> P = np.array([[1 / 2, 1 / 4, 1 / 4], [1 / 3, 1 / 3, 1 / 3], [0, 0,
        ↪  1]])
6
7   ```
8
9   We now see how the vector $\pi$ changes over time:
10
11  ```python
12  >>> for k in range(10):
13  ...         print(absorption.get_long_run_state(pi, k, P))
14  [1. 0. 0.]
15  [0.5  0.25 0.25]
16  [0.33333333 0.20833333 0.45833333]
17  [0.23611111 0.15277778 0.61111111]
18  [0.16898148 0.1099537  0.72106481]
19  [0.12114198 0.0788966  0.79996142]
20  [0.08686986 0.05658436 0.85654578]
21  [0.06229638 0.04057892 0.8971247 ]
22  [0.0446745 0.0291004 0.9262251]
23  [0.03203738 0.02086876 0.94709386]
24
25  ```
```

To test the documentation gives the results that are written, run the following at the command line:

Command line input

```
$ python -m doctest README.md
```

When testing the documentation, if there are no errors there will be no output as shown in Figure 17.3.

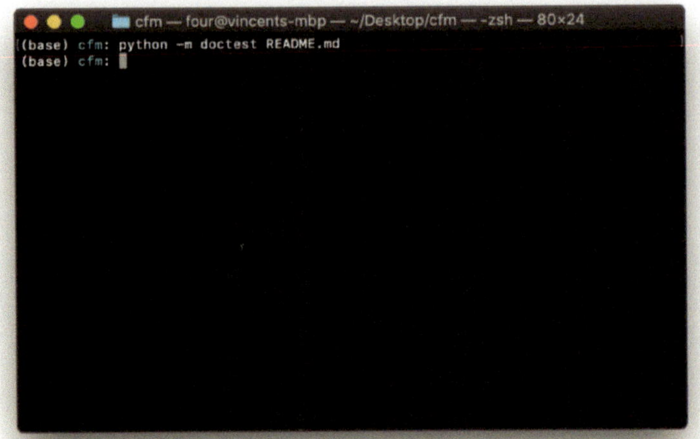

Figure 17.3 Running the doctests with no errors.

Similarly to testing the code, if an error is included in the documentation, an error will be displayed when running the doctests. This is shown in Figure 17.4 where the final output has been changed to include an error: `-1` is written instead of `0.94709386`.

Here is the fully modified tutorial of the documentation:

Markdown input

```
# Absorption

Functionality to study absorbing Markov chains.

## Tutorial

In this tutorial we will see how to use `absorption` to study an
↪ absorbing
Markov chain. For some background information on absorbing Markov
↪ chains we
```

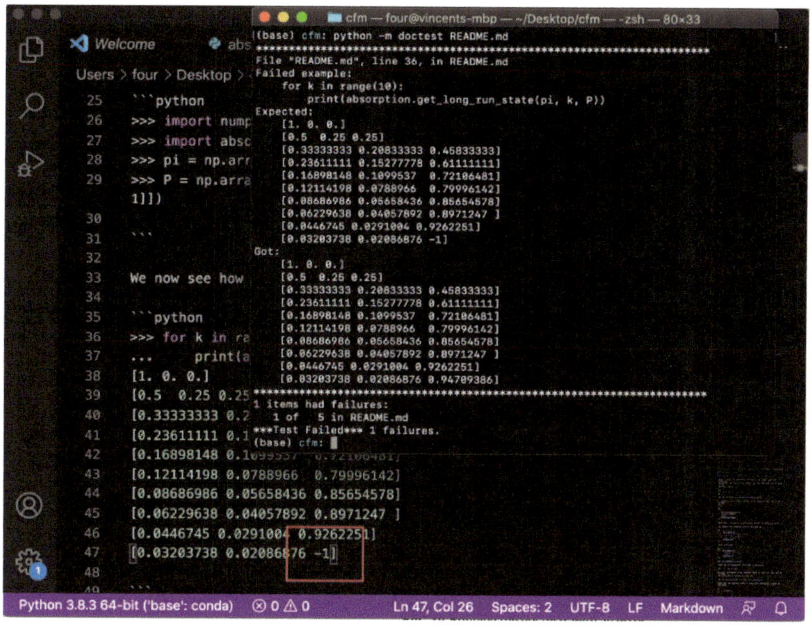

Figure 17.4   Running the doctests with an error.

```
9    recommend: <https://en.wikipedia.org/wiki/Absorbing_Markov_chain>.
10
11   Given a transition matrix $P$ defined by:
12
13   $$
14   p = \begin{pmatrix}
15       1/2 & 1/4 & 1/4\\
16       1/3 & 1/3 & 1/3\\
17       0   & 0   & 1
18        \end{pmatrix}
19   $$
20
21   We will start by seeing how the chain evolves over time by starting
     ↪  with an
22   initial vector $\pi=(1,0,0)$. In the next code snippet we will import
     ↪  the
23   necessary libraries and create both $P$ and $\pi$:
24
25   ```python
26   >>> import numpy as np
27   >>> import absorption
28   >>> pi = np.array([1, 0, 0])
29   >>> P = np.array([[1 / 2, 1 / 4, 1 / 4], [1 / 3, 1 / 3, 1 / 3], [0, 0,
     ↪  1]])
30
```

```
31      ```
32
33      We now see how the vector $\pi$ changes over time:
34
35      ```python
36      >>> for k in range(10):
37      ...         print(absorption.get_long_run_state(pi, k, P))
38      [1. 0. 0.]
39      [0.5  0.25 0.25]
40      [0.33333333 0.20833333 0.45833333]
41      [0.23611111 0.15277778 0.61111111]
42      [0.16898148 0.1099537  0.72106481]
43      [0.12114198 0.0788966  0.79996142]
44      [0.08686986 0.05658436 0.85654578]
45      [0.06229638 0.04057892 0.8971247 ]
46      [0.0446745 0.0291004 0.9262251]
47      [0.03203738 0.02086876 0.94709386]
48
49      ```
50
51      We see that, as expected over time the probability of being in the
        ↪  third state,
52      which is absorbing, increases.
53
54      We can also use `absorption` to get the average number of steps until
55      absorption from each state:
56
57      ```python
58      >>> absorption.compute_t(P)
59      array([3.66666667, 3.33333333])
60
61      ```
62
63      We see that the expected amounts of steps from the first state is
        ↪  slightly more
64      than from the second.
```

17.1.3 Documenting the tests

Finally it is important to document how to run the tests. The **reference** section is an appropriate place to put this. Add the following to the README.md file:

Markdown input

```
1   ### Testing the software
2
3   To test the code:
4
5   ```
6   $ python test_absorption.py
7   ```
8
9   To test the documentation:
10
11  ```
12  $ python -m doctest README.md
13  ```
```

17.2 HOW TO

17.2.1 Write an `assert` statement

An `assert` statement is followed by 2 values: a boolean and an optional string that gets returned if the boolean is `False`.

Usage

```
1   assert boolean, string
```

For example, given a function that adds one to a variable:

Jupyter input

```
1   def add_one(a):
2       """
3       Returns a + 1
4       """
5       return a + 1
```

You can assert the expected behaviour:

Jupyter input

```
1   assert add_one(5) == 6, "The function gave the wrong answer."
```

Note that if you change the function to include an error, for example, here adding 2 and not 1, and run the same assert you get an error as well as the specified string.

Jupyter input

```
1  def add_one(a):
2      """
3      Returns a + 1
4      """
5      return a + 2
6
7
8  assert add_one(5) == 6, "The function gave the wrong answer."
```

```
AssertionError                          Traceback (most recent call last)
Cell In[3], line 8
      2      """
      3      Returns a + 1
      4      """
      5      return a + 2
----> 8 assert add_one(5) == 6, "The function gave the wrong answer."

AssertionError: The function gave the wrong answer.
```

17.2.2   Write `assert` statements for code that acts randomly

When making an assertion about code that behaves in a random manner, use seeding as described in Section 7.2.9.

For example:

Jupyter input

```
1  import random
2
3
4  def roll_a_dice():
5      """
6      Pick a random integer between 1 and 6 (inclusive)
7      """
8      return random.choice(range(1, 7))
```

To test this, include a number of seeded assertions:

Jupyter input

```
1  random.seed(0)
2  assert roll_a_dice() == 4, "The 0 seed did not give the expected
   ↪  result"
3  random.seed(1)
4  assert roll_a_dice() == 2, "The 1 seed did not give the expected
   ↪  result"
5  random.seed(2)
6  assert roll_a_dice() == 1, "The 2 seed did not give the expected
   ↪  result"
7  random.seed(3)
8  assert roll_a_dice() == 2, "The 3 seed did not give the expected
   ↪  result"
```

You can also check behaviour over a number of repetitions:

Jupyter input

```
1  random.seed(0)
2  samples = [roll_a_dice() for repetition in range(1000)]
3  all_values = {1, 2, 3, 4, 5, 6}
4  assert set(samples) == all_values, "Not all values have been obtained
   ↪  over 1000 repetitions"
```

You can also confirm that the count of a given value is as expected:

Jupyter input

```
1  assert [samples.count(k) for k in range(1, 7)] == [193, 150, 166, 170,
   ↪  152, 169], "The count of the values is not giving the expected
   ↪  count"
```

The last assertion used the **count** method on a list that counts a given number of items in a list.

17.2.3 Write a test file

To write tests, assertion statements should be put into a file separate from the code in functions.

For example, if the dice.py file contained:

Python file

```python
1  import random
2
3
4  def roll_a_dice():
5      """
6      Pick a random integer between 1 and 6 (inclusive)
7      """
8      return random.choice(range(1, 7))
```

Then a separate `test_dice.py` file with the following would be written:

Python file

```python
1   import dice
2
3
4   def test_roll_a_dice_with_specific_values():
5       """
6       Check the roll a dice function gives specific numbers for a number
        ↪ of seeds.
7       """
8       random.seed(0)
9       assert dice.roll_a_dice() == 4, "The 0 seed did not give the
        ↪ expected result"
10      random.seed(1)
11      assert dice.roll_a_dice() == 2, "The 1 seed did not give the
        ↪ expected result"
12      random.seed(2)
13      assert dice.roll_a_dice() == 1, "The 2 seed did not give the
        ↪ expected result"
14      random.seed(3)
15      assert dice.roll_a_dice() == 2, "The 3 seed did not give the
        ↪ expected result"
16
17
18  def test_roll_a_dice_for_a_large_sample():
19      """
20      Collect a sample of 1000 rolls and confirm that we have expected
        ↪ results.
21      """
22      random.seed(0)
23      samples = [dice.roll_a_dice() for repetition in range(1000)]
24      all_values = {1, 2, 3, 4, 5, 6}
25      assert set(samples) == all_values, "Not all values have been
        ↪ obtained over 1000 repetitions"
```

```
26      expected_counts = [193, 150, 166, 170, 152, 169]
27      assert [samples.count(k) for k in range(1, 7)] == expected_counts,
   ↪    "The count of the values is not giving the expected count"
28
29  test_roll_a_dice_with_specific_values()
30  test_roll_a_dice_for_a_large_sample()
```

To run the tests you would then type the following at the command line:

Command line input

```
1  $ python test_dice.py
```

17.2.4 Format doctests

When writing code in documentation if you write it using a specific format, then python can be used to confirm that the code and its output is correct.

Markdown input

```
1  >>> <python_code>
2  <expected_output
3  >>> <python_code_over_multiples_lines>
4  ... <python_code_over_multiple_lines>
5  <expected_output>
```

- >>> is marks the start of a python statement.

- ... is used if the statement is multiple lines.

- The expected output is written after the python statements.

For example, if you were documenting the following code written in a `dice.py` file:

Python file

```
1  import random
2
3
4  def roll_a_dice():
5      """
6      Pick a random integer between 1 and 6 (inclusive)
7      """
8      return random.choice(range(1, 7))
```

You would write:

```
Markdown input

1  >>> import dice
2  >>> random.seed(0)
3  >>> dice.roll_a_dice()
4  4
5
```

17.2.5   Run doctests

Given a file with doctests, to run them type the following at the command line:

```
Usage

1  $ python -m doctest <file>
```

For example:

```
Command line input

1  $ python -m doctest README.md
```

17.3   EXERCISES

Write tests for the `statistics.py` file written in the exercises of Chapters 15 and 16.
    Run the tests.

17.4   FURTHER INFORMATION

17.4.1   Why are tests written as functions?

In Section 17.1 you wrote all the tests using `assert` statements inside of functions. **Technically** this is not necessary as you could write a single script with all the assert statements one after the other.

    This is not recommended: by using functions you directly have a place to document the test (in the docstring) and the tests themselves are modularised. Furthermore, this is actually how to write the tests when using a more efficient way of running tests as described in Section 17.4.2.

17.4.2   Is there a more efficient way to run tests?

Writing tests as a script and directly running them has one immediate problem: once the first `assert` statement fails, the rest of them are not run.

    There is a Python library for running tests called `pytest` [7].

17.4.3 What should be tested?

The short answer to this is that all software should be tested and that software is compromised of documentation and code.

Note that it is often not sufficient to test a function in a single way. For example, consider a function that does two different things depending on the parity of some input:

Python file

```python
def feedback_on_guess(guess, chosen_number):
    """
    Returns whether or not a guess is:

    - Larger than  a chosen_number
    - Smaller than a chosen_number
    - Equal to a chosen number
    """
    if guess < chosen_number:
        return "Guess is lower than chosen number"
    if guess > chosen_number:
        return "Guess is larger than chosen number"
    return "Guess is equal to chosen number"
```

In this case you would need to write at least three tests that check the three behaviours. In practice there might be some functionality that is not tested but this should be made clear and explicit. Documentation explaining the lack of tests should be written.

17.4.4 Why do you need doctests?

The purpose of doctests is to ensure that the code written in documentation is correct.

It is important to not use doctests to test the functionality of the code: this risks making the documentation unclear.

17.4.5 What is test driven development?

Test driven development is the development process of writing the test before you write the code. While this might seem counter-intuitive, it is in fact an efficient approach to writing robust code.

In practice the process is as follows:

1. Write a test for some new functionality.

2. Run the tests to confirm that it fails (as the functionality is not yet written).

3. Write the functionality.

4. Run the test.

5. Modify the test and/or the functionality

Steps 4 and 5 might be repeated many times.
A good overview of test driven development is given in [8].

17.4.6   How are modularisation, documentation and testing related?

In Chapters 15, 16, and 17, three concepts that move from writing code that works to writing software that is reliable have been discussed:

- Modularisation.

- Documentation.

- Testing.

In reality **all three** of these concepts are closely related and fundamental to good software. Figure 17.5 shows this.

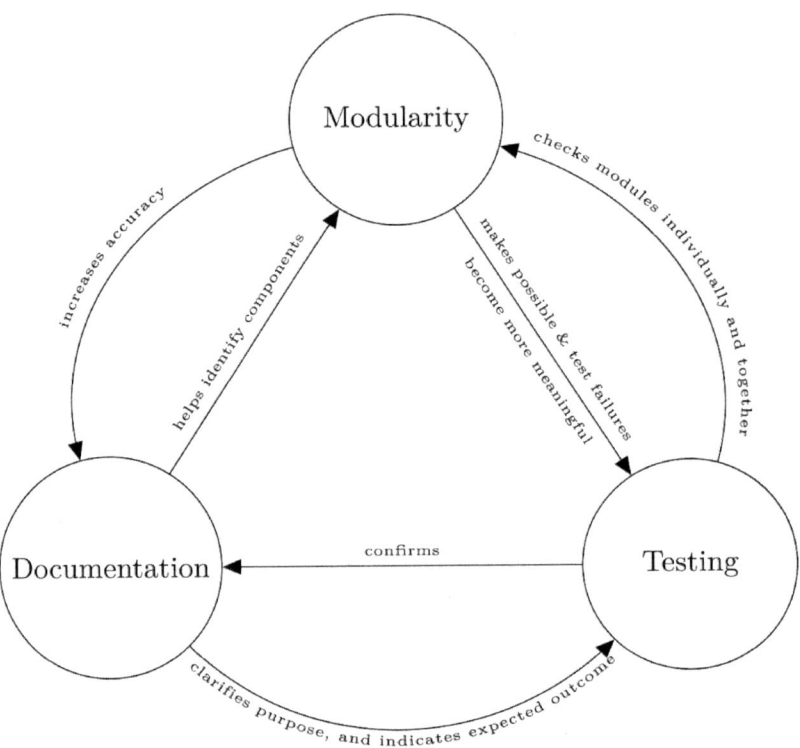

Figure 17.5   The relationship between modularisation, documentation and testing.

IV

About This Book

How This Book Is Written

This chapter include some specific information about the book itself.

This book is written using Jupyterbook [2].

The source files are written in the `myst` format which is a plain text format for Jupyter notebooks. This ensures:

- That the notebooks are version controlled effectively.

- The output of the code is an actual computation.

The computations carried out in this book were carried out with python version `3.10.14`. The library versions used are:

- jupyter version `1.0.0`.

- matching version `1.4.3`.

- matplotlib version `3.9.0`.

- numpy version `1.26.4`.

- scipy version `1.13.0`.

- sympy version `1.12`.

DOI: 10.1201/9781003451860-18

Bibliography

[1] José Carlos Bautista. *Mathematics and Python Programming*. Lulu.com, 2014.

[2] Executable Books Community. Jupyter book, February 2021.

[3] Martin Fowler. *Refactoring: improving the design of existing code*. Addison-Wesley Professional, 2018.

[4] Robert C Martin. *Clean code: a handbook of agile software craftsmanship*. Pearson Education, 2009.

[5] Cyrille Martraire. *Living documentation: continuous knowledge sharing by design*. Addison-Wesley, Boston, 2019.

[6] Sam Morley. *Applying Math with Python: Over 70 practical recipes for solving real-world computational math problems*. Packt Publishing Ltd, 2022.

[7] Bruno Oliveira. *pytest Quick Start Guide: Write better Python code with simple and maintainable tests*. Packt Publishing Ltd, 2018.

[8] Harry Percival. *Test-driven development with Python: obey the testing goat: using Django, Selenium, and JavaScript*. "O'Reilly Media, Inc.", 2014.

[9] Daniele Procida. Diátaxis documentation framework.

[10] Amit Saha. *Doing Math with Python: Use Programming to Explore Algebra, Statistics, Calculus, and More!* No Starch Press, 2015.

[11] Steve Selvin. Monty hall problem. *American Statistician*, 29(3):134–134, 1975.

[12] Guido van Rossum, Barry Warsaw, and Nick Coghlan. Style guide for Python code. PEP 8, 2001.

Index